A Primer of Neuroimmunological Disease

Andrew R. Pachner

A Primer
of Neuroimmunological
Disease

 Springer

Andrew R. Pachner
Department of Neurology and Neuroscience
New Jersey Medical School
University of Medicine and Dentistry of New Jersey
Newark, NJ, USA

ISBN 978-1-4614-2187-0   e-ISBN 978-1-4614-2188-7
DOI 10.1007/978-1-4614-2188-7
Springer New York Dordrecht Heidelberg London

Library of Congress Control Number: 2012930829

Printed on acid-free paper

Springer is part of Springer Science+Business Media (www.springer.com)

Preface

Neuroimmunologists deal with the nervous system and the immune system, both of which are giant, complex communication networks. One would think that we would be excellent communicators ourselves, but we have not communicated well with individuals outside our field. This is very unfortunate because the field of neuroimmunology is more dependent than most on input from fields outside its own. Researchers in basic neuroscience, basic immunology, clinical neurology (to name just a few) have much to contribute and these contributions would be enhanced if those researchers had a better understanding of neuroimmunology. This book is an attempt to provide a basic understanding of some of the neuroimmunological diseases to individuals who are not neuroimmunologists.

Who should use this book? I have lofty ambitions for this book in that I think it can be helpful to a large number of people, such as basic scientists in both neuroscience and immunology, neurologists, and motivated individuals in pharmaceutical companies who are neither PhDs nor MDs. Like Icarus, who had to navigate a path not too close to the sun nor too close to the sea, I tried to be basic enough to be understood by neophytes yet having enough depth so that it would not lose the attention of more-educated readers. Time will tell whether I succeeded.

Why now? There are many reasons. One reason is that the most common neuroimmunological disease, multiple sclerosis, covered in five chapters in this book, is attracting increasing interest from pharmaceutical companies as a therapeutic target. "Unmet need" is an understatement when it comes to improving our therapy of this prevalent disabling disease of the young. Another reason for the need now for this book is that there has been a paradigm shift in our understanding of the immune response within the nervous system. Far from being an "immune-privileged" tissue, haughtily excluding itself from any immune functions, the nervous system is actively involved in immune responses; it is simply that it participates using its own rules which are actively being researched but remain to be fully elucidated. A third reason for the need for this book is that increasingly, diseases thought to have a "degenerative" etiology, such as stroke or Alzheimer's, have a component related to the immune system. This subdivision within neuroimmunology, discussed briefly in Chap. 15, will likely grow substantially in the near future.

How could I possibly cover neurology, immunology, neuroimmunology, and neuroimmunological disease in a short book? The short answer is that I can't.

However, this is a primer, not a complete textbook. It will help the reader most as an introduction, and as a guide to what areas to pursue in the literature. It is customary for an author to lament about how large his subject is, and how many corners needed to be cut, and I will certainly adhere to the custom. But I tried to make the field of neuroimmunology understandable to a wide audience without being too lengthy.

Have I been too cynical about the state of our knowledge and the efficacy of available therapy? One of the physicians in the field who reviewed some of the chapters thought so. He felt I was "riding roughshod" and being too "curmudgeonly" and wished me to "explain the various drugs with great enthusiasm." I apologize to him, and to those who wanted this book to be more upbeat. I did not intend to have any part of the book interpreted in a way that in any way is negative about the field of neuroimmunology or those practicing this subspecialty. I have been a neuroimmunologist for 30 years and love the field and the people who work in it. However, I also adhere to the tenet, "primum non nocere" (the first rule is to do no harm), and I feel that the benefit/risk considerations should be clearly weighted toward benefit prior to recommending a therapy. Unfortunately, the trend in the field is to move in the other direction, toward therapies that are increasingly risky with questionable benefit to show for it. It is possible that in a future of evidence-based medicine and increasing accountability, we will have better tools to measure benefit/risk ratios in order to avoid major side effects and to maximize benefit.

What's in the future for neuroimmunology? I see the partnership between basic neuroscience and neuroimmunology becoming stronger, and advances in our understanding leading to further major advances in diagnosis and therapy. We will benefit from advances toward neuroprotective therapies in other parts of neuroscience to provide clues to ameliorate neurodegeneration in neuroimmunological diseases. Ultimately, our understanding of MS will increase and we will identify more and more effective therapies. From my mouth to God's ears…

I could not have written this book without a great deal of help. Steve Kamin, the chairman of our department of neurology at UMDNJ—New Jersey Medical School, was very supportive and allowed me to take sabbatical time. Susan Goelz, Lew Fredane, David Lagunoff, Norm Kachuk, Steve Kamin, and Stuart Selonick edited chapters, and aligned my frequently muddled efforts. The staff at Springer were extremely helpful, especially Andy Kwan and Richard Lansing. My daughter, Anna, helped considerably with image issues. And of course my long-suffering wife, Barbara, who had to put up with my periods of both mania and depression, was always there for emotional support.

Newark, NJ, USA   Andrew R. Pachner

Acknowledgments

Acknowledgments for Helpful Discussions and Providing Material

Special thanks to Susan Goelz, Lew Fredane, Norm Kachuk, Steve Kamin, David Lagunoff, and Stuart Selonick who patiently read through the tortured prose of early versions and made much-needed recommendations.

Jack Antel
Klaus Bendtzen
Joe Berger
Bruce Cohen
Nicolas Collongues
Kathy Conant
Gary Cutter
Martin Daumer
Peter Dyck
Florian Eichler
Patricia Fitzgerald-Bocarsley
John Foley
Doug Green
Ken Gorson
Wayne Hogrefe
Doug Jeffrey
Dimetrios Karussis
Susumu Kusunoki
Norman Latov
Hans Lassman
Vanda Lennon
Howard Lipton
Bob Lisak
Michael Lockshin
Christina Marra
Jennifer Michaels
Jana Preiningerova
Harry Prince
Kotil Rammohan
John Richert
David Richman

Moses Rodriguez
Myrna Rosenfeld
Walter Royal
Subraminam Sriram
Israel Steiner
Carlo Tornatore
Helen Tremlett
Ken Tyler
Angela Vincent
Brian Weinshenker
Hugh Willison
Gil Wolfe
Robert Yu

Special thanks to my wife, Barbara, who patiently tolerated my idiosyncrasies, and to my daughter, Anna, who assisted me with getting images ready for the book. Thanks also to Richard Lansing, my editor, who made my first experience as an author of a single-author text a pleasant one.

Contents

1 The Beginnings of Immunology

The focus of immunology, both historically and conceptually, is the defensive response of a vertebrate animal (i.e., humans or animal models) to microorganisms leading to clearance of the pathogen, maintenance of homeostasis, and enhanced protection in the future to the same or similar microorganisms. Prior to the second half of the twentieth century, infectious diseases, especially those encountered during childhood, were massive public health threats. Although vaccines and antibiotics have blunted the threat, infections remain major causes of morbidity and mortality, and the immune response to established infections, as well as new/emerging infections, remains the nexus of immunology. An example of the importance of infections in immunology is that in many medical schools, microbiology and immunology are within the same department.

The father of immunology is generally felt to be Edward Jenner, who, at the end of the eighteenth century began using cowpox vaccination in England to protect against smallpox (see Inset 1.1 and Fig. 1.1). However, he was not the first person to utilize immunology to protect against infection. The practice of "variolation," purposeful inoculation with smallpox (Variola) in a controlled manner to limit the effects of natural infection, had been around for many decades and possibly for centuries prior to Jenner. Smallpox had been a major public health problem in civilization for many centuries.

Inset 1.1 The Case of Smallpox and the Power of Immunology

At-risk population. All ages in all countries prior to 1950. Hundreds of thousands of Europeans died each year from the infection.

Cause. Infection with the smallpox (variola virus), a very large virus of the poxvirus group, consisting of a genome of almost 200,000 bp of double-stranded DNA packaged into a virion measuring about 200×300 nm.

Symptoms and signs. After an incubation period of about a week and a half, fever, malaise, and a rash appear. The rash goes through a predictable cycle of papules, vesicles, and pustules, which ultimately crust (see Fig. 1.1). Transmission is via the respiratory tract followed by periods of viremia resulting in spread to the rest of the body.

Morbidity/mortality. Mortality of the acute infection was generally around 30% of those infected, although with children, the number was higher. Individuals recovering from smallpox usually experienced no sequelae except for pockmarks on the face which could be disfiguring.

Treatment. None.

Prevention. Vaccination. Intensive global vaccination efforts led to the global eradication of smallpox so that by 1980, smallpox was declared as being eradicated from the world.

A.R. Pachner, *A Primer of Neuroimmunological Disease*,
DOI 10.1007/978-1-4614-2188-7_1, © Springer Science+Business Media, LLC 2012

Fig. 1.1 A child with smallpox. This girl from Bangladesh developed smallpox in 1973. Freedom from smallpox was declared in Bangladesh in December, 1977 when a WHO International Commission officially certified that smallpox had been eradicated from that country. From CDC image files available in the public domain at http://commons.wikimedia.org/wiki/File:Child_with_Smallpox_Bangladesh.jpg)

However, as the field of immunology has matured, it has begun to be applied to many more fields outside of microbiology and infectious diseases. For instance, an increasing number of diseases are being recognized as being "autoimmune" in which the normal barriers to immune reactions against self-components are breached. Also there are diseases that are mistakenly characterized as being "autoimmune"; the definition of autoimmunity within neuroimmunology will be discussed in Chap. 3. Another area of intense interest within immunology has been the immune response to cancers, since immunosuppression either via medications or via infections such as the human immunodeficiency virus (HIV) results in increased incidence of malignancies. And even more recently, within the past decade, neurological diseases, thought previously to be "degenerative," such as stroke, Parkinson's, and Alzheimer's disease, (see Chap. 15), have been found to have substantial immunological contributions.

2   The Components of the Healthy Immune Response

Given the fact that the historical foundations of immunology were laid in man's response to infection, and infectious diseases are a battle between the pathogen and the host's immune system, it is not surprising that much of the language of immunology is the language of war. There is "enemy," "invasion," "targeting," "killing," "defense," "decoy," etc. A human's immune system must be extremely powerful to overwhelm the large variety of enemies invading the body, and ultimately the power of the immune system must be greater than the power of the invader, or the invader will kill the human. Just as a prosperous, civilized society such as the USA, or a highly functional animal such as the human, must have the power of a strong military/immune response, both must also have a very complex, sophisticated regulatory mechanism, so that the power is not brought against the state/human itself. Both the human and the state suffer terrible consequences if the immune/military arm is inadequately or excessively controlled, and both constantly adjust the regulatory mechanisms to achieve optimal balance. Unfortunately, our understanding of both the weapons and control of the immune system are primitive, and our attempts at manipulating the immune response in disease are thus necessarily crude, as will be discussed in Chap. 19.

Two helpful classifications of the immune response are those of *cells/molecules* involved, summarized in Table 1.1, and their participation in the *innate vs. adaptive* immune response. The *innate immune response* occurs early after infection begins. Thus, in the example of Lyme disease, an infection that is clinically important in many parts of the USA and Europe, the innate immune response occurs within the first few days after the tick bite which injects the causative bacterium, *Borrelia burgdorferi*, into the skin.

Table 1.1 Major cells/molecules involved in the healthy immune response

Cell/molecule	Subclassification
White blood cell	Lymphocytes, polymorphonuclear leukocytes, macrophages, and monocytes, dendritic cells
Immunoglobulin	Isotypes
Major histocompatibility complex (MHC)	I and II
Cytokine	Cell derivation, TNF superfamily, chemokines, interferons
Cluster of differentiation (CD) antigens	Surface proteins used to identify functionally distinct immune cells

Polymorphonuclear leukocytes and macrophages are quickly recruited to the site of infection leading to redness, swelling, and local pain. Natural antibodies, which are present without previous exposure to the pathogen, bind to the bacteria and activate complement. Cytokines, important communication and effector molecules of the immune response (see below), accumulate locally and enhance the inflammation. Products of degradation of the bacteria activate a variety of cells through Toll-like receptors (TLRs). All of these processes utilize weapons against the bacteria that are already present and ready, and none require the production of novel molecules, or the generation of new classes of cells.

In contrast, by 5–7 days after infection, many of these processes are waning, and the *adaptive immune response* is taking over. The adaptive immune response, led by B cells and T cells with receptors specific for *Borrelia burgdorferi* proteins, is necessarily later than the innate response, since it utilizes molecules and cells specifically "adapted" for the invading organism, which are byproducts of relatively complex and thus slower processes, such as genetic recombination and hypermutation occurring in lymphoid tissue. These cells, once "produced" and replicating, traffic to the sites of infection and local lymphoid tissue, utilizing adhesion molecules to guide them, and continue to proliferate. Lymphocytes which do not have specificity for the pathogen's antigens also traffic to sites of infection, but do not proliferate, and instead, without stimuli to proliferate, die in an orderly regulated process of death called apoptosis. The selective expansion of relevant cells and death and clearance of unneeded lymphocytes is an important part of the immune response.

Apoptosis, also called programmed cell death (PCD), which has only attracted interest in the last few decades, is a critical process in a highly regulated system such as the immune response. It is induced by both intracellular and extracellular signals, which can be either pro- or anti-apoptotic. The final effectors are proteolytic cysteine-dependent aspartate-directed proteases (caspases). Important apoptotic molecules involved in the final regulation of caspases are the cytokine tumor necrosis factor (TNF) and its receptors, Fas receptor (also called CD95) and Fas ligand, members of the Bcl-2 family.

If the infection persists, the weapons of the adaptive response become more and more powerful, and in the vast majority of infected individuals the power of the immune response is adequate to kill all the spirochetes and the infection is cleared. Only a small percentage of unfortunate infected humans become persistently infected with the spirochete, and are at risk for developing some of the more serious late complications such as neurological complications or arthritis. In Lyme disease as in most infections, the factors that determine why a relatively small percentage of infected individuals cannot clear the infection are unknown and likely relate to both host and pathogen factors.

Key components of the adaptive immune receptor are a family of receptors on the surface of B- and T cells, called antigen receptors, which bind with high affinity to the newly identified antigens. The B-cell receptor (BCR) consists of the Fab of the immunoglobulin molecule produced by the B cell (see discussion of immunoglobulin structure below) and can bind the antigen directly without the need for another cell type. Activation of the T-cell receptor (TCR) requires an antigen-presenting cell (APC) which "processes" an antigen partly by chopping it into small pieces, and then presenting it to the T cell after it has bound to the major histocompatibility complex (MHC) complex. This combination of the APC and its MHC complex, the antigen piece, and the TCR is sometimes called the "trimolecular complex" of T-cell activation.

Once the bacteria have been cleared, the immune system of the individual infected with *Borrelia burgdorferi* does not "forget" the infection, but utilizes one of the key aspects of the adaptive immune response, i.e., immunological memory. Memory B cells, memory T cells, and immunoglobulins which bind with high affinity to the spirochete all continue to be present for years after infection, possibly for the lifetime of the individual, and protect from future infection. These powerful weapons are dormant but can leap to action should any infection resembling Lyme disease return.

2.1   White Blood Cells

Centrifugation of blood separates cells in the blood into two cellular compartments, the red compartment comprising red blood cells, and the white compartment, consisting of the less numerous, but much larger and more complex, white blood cells. White blood cells are initially produced from progenitor cells in the primary lymphoid organs, the bone marrow and thymus, circulate throughout the body, mostly into peripheral lymphoid organs such as the lymph nodes, mucosa-associated lymphoid tissue (MALT), and the spleen. Some white blood cells, such as plasma cells, terminally differentiated B-lymphocytes which produce antibodies, circulate back into the bone marrow.

2.1.1   Lymphocytes

Lymphocytes are the most well-studied of the white blood cells. The National Library of Medicine data base records 480,000 scientific articles on lymphocytes in the past 50 years, far more than any other of the immune cell types. Lymphocytes can be either small, 7–10 μm in diameter, such as the thymus-derived cell (T cell) or the bursa-derived cell (B cell), or large, 10–13 μm in diameter, such as natural killer cell (NK cell). There are approximately 500 billion, i.e., 5×10^{11}, lymphocytes present in the human body at any one time, almost all of them in the lymphoid tissues. Lymphocyte kinetics are quite complex for two reasons. First, lymphocytes are travelers, trafficking constantly through lymphoid tissues and blood. Secondly, they frequently die soon after coming to maturity only to be replaced by newly produced lymphocytes. The movement of lymphocytes in response to a stimulus to the immune system, including a brief description of lymph nodes, lymph vessels, and trafficking to the nervous system, is in Chap. 3.

B Cell

The "B" in B cell comes from the name "bursa of Fabricius," a specialized organ in birds; this sub-population was first described in the 1960s as "bursa-derived" in birds. Later research determined that mammalian B cells develop in the bone marrow and spleen. B cells have many functions, but most immunologists would consider their primary role to be to develop into immunoglobulin-producing cells, called plasma cells, responsible for the immunoglobulins, also called antibodies, which circulate and are involved in many effector and protective functions of the immune system. Most B cells, however, never develop into plasma cells, and B cells which precede plasma cells in development can function by producing cytokines and by interacting with other immune cells. How important these multiple roles are in most immune responses is not well understood. For example, the role of B cells in the inflammatory activity of multiple sclerosis was unanticipated, until the strong anti-inflammatory effects of B-cell-depleting therapy were identified in a clinical trial treating MS with the anti-CD20 monoclonal antibody rituximab (see Chaps. 17 and 19). However, for survival purposes, the primary role of B cells in humans is the production of antibody since children born without functional B cells, such as in X-linked agammaglobulinemia (XLA) (see Inset 1.2) die of infections unless treated with intravenous immunoglobulins; many children so treated will live nearly normal life spans because of the strong protective effects which immunoglobulins exert against infection.

Mature B cells are unique among other cells by surface expression of the BCR, which resembles an antibody molecule inserted into the membrane with the antigen-binding portion of the molecule on the outside. The antigen-binding portion, consisting of IgD or IgM antibodies, is bound to the surface. After binding and during cell activation, the BCRs cluster together. The characteristics of the BCR and its cousin, the TCR, are summarized in Table 1.2. Most, but not all, B cells also have the surface expression of CD19 and CD20.

B cells can be further divided into B-1 and B-2 cells. B-1 cells are present in the fetus, and self-renew outside of lymphoid organs, while B-2 cells are produced after birth in the bone marrow and spleen. B-1 cells, most of which express the marker CD5, are responsible for the production of IgM and IgG3 natural antibodies which are specific for more than one antigen, and which have low affinities of binding. B-2 cells, in contrast, produce IgM and IgG with relatively high affinity which require more extensive rearrangement of immunoglobulin molecules called somatic hypermutation. B-1 cells can be purified from peritoneal cavity lavages, while B-2 cells can be isolated from peripheral blood, bone marrow, or spleen.

T Cell

T cells are defined as small lymphocytes, which bear surface TCR complexes, and do not have immunoglobulin on their surface. T cells are the most common lymphocyte type in peripheral blood and have many roles including direct killing of cells infected with viruses, enhancing antibody production by B cells, secretion of cytokines, and regulation of immune function. The "T" in T cell comes from the thymus, the organ located beneath the sternum where T cells mature.

Table 1.2 Characteristics of antigen-specific receptors

Cell	Presence of antigen-specific receptors	Can bind antigen without involvement of other cell type	Requires presence of "antigen-presenting cell"	Minimum requirements for protein antigen
B cell	Yes	Yes	No	Conformational structure
T cell	Yes	No	Yes	Peptides
Other white cells	No	No antigen-binding receptors	No antigen-binding receptors	No antigen-binding receptors

Early in life, the thymus is responsible for generating large numbers of T cells many of which die from positive or negative selection. Through adulthood, the thymus atrophies at a steady rate, and new T cells are generated from expansion of already existing T cells. The surface marker CD3, a critical part of the TCR, is often used to identify T cells. T cells were first described in the 1960s by Jacques Miller and others who revised earlier conceptions of the thymus as a vestigial organ. Since their discovery, T cells have been the most studied of the lymphocytes accounting for nearly 300,000 scientific articles listed in the National Library of Medicine data base. T cells are critical for survival; HIV infection kills infected humans by depletion of T cells (see Inset 1.3).

Inset 1.3 HIV/AIDS and Opportunistic Infections from T-Cell Deficiency

At-risk population. Adults in all countries, but particularly those practicing high-risk activities, such as intravenous drug abuse and promiscuous sexual intercourse. Can be transmitted to infants by infected mothers.

Cause. Infection with the HIV, which is a lentivirus, a member of the retrovirus family, of about 10,000 bases, and transmitted as a single-stranded, positive-sense, enveloped RNA virus. Disease and death in AIDS, the final stage of HIV infection, are generally due to opportunistic infections caused by HIV-induced depletion of CD4+ T cells; dysfunction of NK cells is also thought to be important in susceptibility to OIs.

Symptoms and signs. There are four stages to the infection. The first is an incubation period of approximately a month after initial infection. The second is an "acute infection" period also of about a month of a nonspecific viral syndrome of fever, malaise, swollen lymph nodes, and muscle aches. The third is a latent, asymptomatic phase.

Inset 1.3 (continued)

The mean duration of this phase in an untreated individual is about 10 years. The final stage is that of AIDS, the development of opportunistic infections (OIs). Both CD4+ T-cell counts and viral load in the blood are used as important biomarkers in the disease.

Morbidity/mortality. Death in untreated individuals usually occurs within a year after the development of AIDS, the final stage.

Treatment. Highly active antiretroviral therapy (HAART). A cocktail of a variety of anti-viral drugs which keeps the viral load low and leads to prolonged remission.

Prevention. Avoidance of exposure to the virus is the only way at this time to prevent HIV infection. At this time no vaccine has demonstrated major efficacy.

T cells are commonly divided into "helper" T (T_h) cells which have CD4 on their surface and "help" B cells make antibody, "cytotoxic" T (T_c) cells, which have CD8 on their surface and kill virally infected cells, and "regulatory" T cells which do not have a clear-cut CD signature. T_h cells can be further differentiated by the types of cytokines they produce: Th1 produces mostly macrophage-activating cytokines such as IFN-γ and TNF-β; Th2 produces primarily B-cell-activating cytokines such as IL-4, IL-5, and IL-10; and Th17 cells produce IL-17 and IL-6. Although regulatory T cells are considered by some to be relative newcomers in immunological research, they were actually described in the 1970s by Richard Gershon [1] (Fig. 1.2), an immunologist at Yale, and were termed "suppressor" T cells [2]. "Suppressor" T cells, now called $T_{regulatory}$ cells, or T_{reg} cells, have strong effects on immune processes, but their molecular and phenotypic characterization is extremely difficult, and they remain cells of mystery [3, 4]. The fate of many T cells is to circulate and die relatively quickly, but others become "memory" T cells and

Fig. 1.2 Richard K. Gershon, an immunologist at Yale who was ahead of his time when he first described regulatory T cells, which he called suppressor T cells, in the early 1970s. This image is used with permission of Douglas Green, Ph.D.

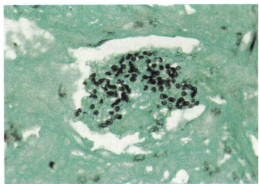

Fig. 1.3 *Pneumocystis jirovecii* detected by methenamine silver stain from the sputum of an HIV patient with *Pneumocystis pneumonia*

can remain viable for many years without known activation; these T cells, as well as their memory B-cell cousins, are the basis for immunological memory, in which exposure to a previously encountered antigen results in a highly accelerated response relative to the first exposure.

Not surprisingly for an entity as powerful as the immune system, tight regulation is required. T_{regs} are only one of a number of tools the immune system uses in self-regulation. Another well-studied down-regulatory mechanism is tolerance, in which cells that normally should be able to respond to an antigen are unable to, due to a poorly understood state of paralysis of the cell. Thus, the normal individual has circulating immune cells which are reactive to various self-antigens, but these are tolerized and do not respond pathologically. Some investigators feel that it is loss of self-tolerance that is the cause of many autoimmune diseases.

Individuals with T-cell deficiencies (see Inset 1.3) develop opportunistic infections (OIs), such as *Pneumocystic pneumonias* (Fig. 1.3), and have an increased risk of malignancies, such as Epstein–Barr virus-associated lymphoma and human herpes virus 8-associated Kaposi's sarcoma. The most common OIs in AIDS patients are those with pneumocystis (a fungus), Candida (a fungus), tuberculosis (a bacterium), and cytomegalovirus. In patients with HIV/AIDS and neurological symptoms, other important OIs are CNS toxoplasmosis and progressive multifocal

leukoencephalopathy (PML) caused by infection in the brain with the JC virus, a polyomavirus.

T- and B cells which appear in the blood or ultimately in areas of inflammation within the nervous system in neuroimmunological disease represent the survivors of a process of the normal extensive loss of lymphocytes through apoptosis, or PCD; these surviving lymphocytes meet complex "quality control" criteria. This selection process occurs in secondary lymphoid tissue such as the lymph nodes and spleen. Survivors of this process can be considered as having been "rescued" from aptotosis, and once these lymphocytes become mature, most of them ultimately die unless they are activated. Activation occurs via interaction with their antigen and appropriate other signals; part of the activation also involves proliferation of the lymphocyte resulting in considerable expansion of the clonal population. Proliferation of lymphocytes ex vivo can also be stimulated by mitogens, proteins which nonspecifically activate lymphocytes (see discussion of lectins in Chap. 18). The progeny of these activated lymphocytes include "memory" cells, which can last for years, and are the basis of immunological memory.

2.1.2 Polymorphonuclear Leukocytes (Neutrophils)

Neutrophils, which constitute about half of the total circulating white blood cells, can leave the circulation very quickly after the beginning of an infection and enter tissues, and represent one of the earliest weapons in the innate immune response. Neutrophils phagocytose bacteria and other pathogens which have been coated with antibodies and complement, and use their granules, which contain a variety of compounds, to kill pathogens.

2.1.3 Monocytes and Macrophages

Monocytes circulate in the bloodstream for about 1–3 days after which they migrate into tissues where they become tissue macrophages. They phagocytose, i.e., ingest, foreign substances such as bacteria. This can be done directly by the recognition of certain nonhuman molecular patterns on microbes by the macrophages. The microbial structures recognized by the innate immune system are sometimes referred to as pathogen-associated molecular patterns (PAMPs) and the receptors for them present on macrophages are called pattern recognition receptors (PRRs). Bacterial lipopolysaccharides and flagellar proteins contain PAMPs which bind to specialized receptors on macrophages and dendritic cells called TLRs, a common class of PRRs. Macrophages also have receptors for the Fc portion of antibody and thus can be activated to kill microbes which are coated by antibody and complement, a process called antibody-dependent cellular cytotoxicity (ADCC).

An important molecule that is upregulated in macrophages during an immune response is nitric oxide (NO), a gas produced by the enzyme inducible nitric oxide synthetase (iNOS). NO is felt to be an important mediator in both immune and nonimmune functions, but its precise role is unclear.

Another cell type of the innate immune response that participates in ADCC via their surface expression of Fc receptors, and resembles macrophages in some ways, are NK cells, natural killer cells, which are large and granular; NK cells are considered key parts of the innate immune response. NK cells are of lymphocyte, not monocyte lineage; they are derived from the same lymphoid precursors as those which give rise to B- and T-lymphocytes. They are negative for the expression of CD3, the canonical T-cell marker, and produce IL-10 and other cytokines without the need for prior stimulation. Some NK cells express CD56 at high levels. These CD56^{hi} cells may have immunoregulatory functions and are thought to be important in the mechanism of action of daclizumab, a drug in phase 3 studies of multiple sclerosis. Interestingly, CD56 is neural cell adhesion molecule (NCAM), which is also expressed on the neuronal and glial surface, and is thought to be important in CNS function.

Another important cell type related to macrophages and derived from monocytes are dendritic cells (DCs). DCs are rarer than macrophages, but are critical links between the innate and adaptive immune response. Immature DCs sample their environment through the extension outward of multiple "dendrites" and become mature when they are activated by an antigen, at which point they migrate to a local lymph node. There they activate antigen-specific T cells, part of the adaptive immune response. Mature DCs are considered the best cells for antigen presentation to T cells since they can activate naive as well as memory T cells, while B cells and macrophages, which can also present antigen to T cells, can activate only memory T cells. There are two main types of human DCs: myeloid dendritic cells (mDCs), which secrete IL-12, and plasmacytoid dendritic cells (pDCs), which secrete type I interferons, mostly interferon-α. The term "dendritic cell" has also confusingly been given to an unrelated "follicular dendritic cell" (Fig. 1.4), a macrophage-like residential cell of lymphoid tissue, which is of mesenchymal, not hematopoietic, origin, and does not express MHC class II antigens, and also has prominent dendrites.

2.2 Molecules

2.2.1 Immunoglobulin

Immunoglobulins (Igs) are products of plasma cells, which are terminally differentiated B cells. They are very large molecules (Fig. 1.5), ranging from 150 kDa for IgG to 970 kDa for IgM; in contrast, cytokines such as chemokines or interferons (described below) are 12–20 kDa. The prototypic effector molecules of the adaptive immune

Fig. 1.4 Scanning electron microscopy (SEM) of a follicular dendritic cell showing dendritic regeneration

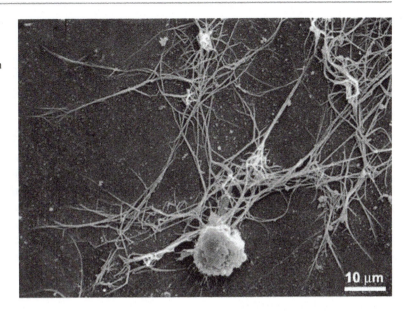

Fig. 1.5 Immunoglobulin molecular structure. The immunoglobulin molecule is composed of two heavy chains and two light chains. Light chains exist in two classes: kappa and lambda. Either type may associate with any type of heavy chain isotype

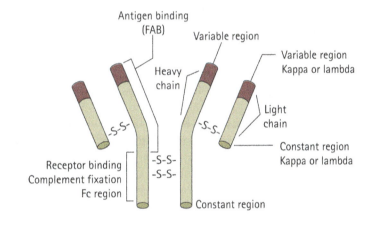

response are immunoglobulins of the IgG isotype, molecules with high affinity for their epitopes which begin to appear in the second week of an immune response against a pathogen. Each single clone of plasma cells, produces many copies of an immunoglobulin unique for that plasma cell, and many plasma cell clones contribute to an effective adaptive antibody response. The differences in the amino acid sequence of the immunoglobulins produced by one plasma cell relative to those produced by other plasma cells can be divided into differences in the Fab or Fc portion; "ab" is an abbreviation for "antigen binding" and "c" is an abbreviation for "constant." The differences within the Fab portions of immunoglobulin molecules, clustered in the first 110 amino acids of the amino terminal ends of the heavy and light chains (colored red in the diagram), is dramatic since there are "hypervariable" areas within Fab that determine the ability of these molecules to bind antigenic epitopes. The variable region of the Fab represents the end product of a molecular restructuring tour de force, a product of a combination of germ line sequences, recombination, somatic hypermutation, gene conversion, and clonal deletion which leads to a very impressive product, whose affinity of Fab for antigen can be as high as 10^{-12} M, i.e., nearly irreversible binding. Thus, due to all of this modification, the final 110 amino acid sequence of the molecule bears little relation to the encoded Fab sequence in the germ line. The Fc portion is much less variable as identified in its name, i.e., the "constant" region, which has no role in antigen binding but determines the interaction of immunoglobulins with macrophages and complement. There are five major classes or isotypes determined by the structure of the Fc portion of the heavy chain: IgG, IgA, IgM, IgD, and IgE, in order of concentration in the serum. IgG is by far the most abundant and has four subclasses.

Immunoglobulins also participate in the innate immune response. B1 B cells produce immunoglobulins which are polyspecific with relatively low affinity for their antigen, called "natural" antibodies. These molecules are not the product of the extensive genetic manipulations described above and are encoded from germline sequences, and circulate at all times ready for interaction with any pathogen that might enter the body. Many natural antibodies are specific for the sugar galactose and other carbohydrate antigens. These natural antibodies, consisting primarily of IgM and IgG3, which are strong activators of complement, do not require prior exposure to any pathogens; many of them recognize epitopes common to bacteria. Their binding to bacteria, and local activation of complement, allows the local participation of granulocytes and other cells and effector molecules of the innate immune response.

2.2.2   The Major Histocompatibility Complex

This is an exceptionally polymorphic gene complex on chromosome 6 in humans which contains over 200 genes; most other mammals have similar numbers of genes in their MHC, while chickens only have 19 genes. Its survival value likely derives from its ability to efficiently present peptides from pathogens processed by APCs to the appropriate T-cell populations; in addition, a number of other cell types other than T cells are also dependent upon the MHC for their function and development. The largest classes of MHC genes are class I and II, distinguished by the types of peptides which they are able to present. In humans, MHC class I genes are divided into HLA-A, -B, and -C, while class II genes are HLA-DR, -DP, and -DQ. The most polymorphic of these genes is the beta subunit of the HLA-DR molecules which has over 500 different alleles.

The recognition of antigens by antigen-specific T cells requires self-MHC. If the same antigen is presented to an individual by cells which either do not bear MHC or bear MHC different than that of the individual, no activation will take place. The Nobel Prize for Medicine was given in 1996 to Peter Doherty and Rolf Zinkernagel for the demonstration that this "MHC restriction" was an invariant feature of activation of murine cytotoxic T cells in the killing of the lymphocytic choriomeningitis virus. The precise molecular mechanisms by which this MHC restriction is under active investigation, but it was clear that in the adaptive immune response activation of specific T cells must be in the "context" of MHC molecules, i.e., require processing of antigen by APCs and presentation of antigenic peptides to T cells on the MHC molecule. Another clinically important role of MHC molecules is their importance in transplantation since the smaller the difference in MHC between a transplant and the recipient, the less the likelihood of transplant rejection. Major problems with the MHC result in immunodeficiency syndromes, one of which is the "bare lymphocyte syndrome" (see Inset 1.4).

Inset 1.4 MHC and Immune Deficiency—Bare Lymphocyte Syndrome (BLS)

At-risk population. Rare genetic disease.

Cause. This is caused by a variety of mutations which result in the absence of MHC molecules, usually caused by mutations in genes outside of the MHC, frequently critical transcription factors. One of these is class II transactivator (CIITA). BLS falls under a classification of immune deficiencies called severe combined immunodeficiency (SCID).

Symptoms and signs. There is extreme susceptibility to viral, bacterial, and fungal infections. Thus, the usual presentation is an infant or toddler with frequent, severe infections.

Morbidity/mortality. High. Death from opportunistic infections is common.

Treatment. Bone marrow transplantation is the only curative treatment. Supportive treatment with IV immunoglobulin and chronic treatment for common infections such as pneumocystis are used.

2.2.3 Cytokines

Cytokines are small-to-moderate-sized proteins which serve as communication molecules between cells; they usually operate at very short range. One of the first groups of cytokines described were interleukins, with IL-1 having one of the longest histories [5]. This cytokine has many roles and was described in the 1950s as one of the "endogenous pyrogens," substances that could be transferred from one animal to another to cause fever. Cytokines generally are secreted by cells after activation and impact on other cells via cytokine-specific receptors. The field of cytokines has experienced explosive growth in the past 30 years, and now cytokine antagonists or recombinant cytokines are increasingly used in therapy of a wide range of diseases; the biologic therapy most commonly used in treating neuroimmunological disease is the cytokine, interferon-β (see Chap. 7).

Cytokines are generally difficult to categorize, but large groups include:
- Interleukins, which act primarily upon white cells
- Interferons, which "interfere" with viral replication in cells targeted by these cytokines
- Members of the TNF superfamily, and
- Hematopoietic growth factors.

Interferons will be discussed at length in Chap. 7. The TNF superfamily has been targeted extensively by multiple products in the treatment of rheumatoid arthritis (see Inset 1.5 and Fig. 1.6). Cytokines function by binding to specific receptors present on their cellular targets. Cytokine function then is dependent not only on concentration of the cytokine in the extracellular space but also the concentration and proper function of the cytokine receptors on the target cell. An example of the interaction of a cytokine with its

Inset 1.5 Rheumatoid Arthritis; Treatment with Cytokine Blockers

At-risk population. Very common disease, afflicting 1% of the general population, women more frequently than men.

Cause. The cause of rheumatoid arthritis is unknown; it results in progression inflammation and damage to joints throughout the body.

Symptoms and signs. Patients have pain, swelling, and limitation of movement of joints throughout their body.

Morbidity/mortality. The pain and decreased use of RA can be highly disabling.

Treatment. The mainstay of therapy for many years has been immunosuppressives such as methotrexate. Recently, therapies which target the proinflammatory cytokine TNF-α have been developed. They include infliximab, adalimumab, certolizumab, and golimumab which are monoclonal antibodies specific for TNF-α and etanercept, which is a fusion protein of a soluble TNF receptor linked to human IgG1 Fc.

Fig. 1.6 Rheumatoid arthritis. This disease is a crippling, painful disease of joints in which drugs targeting the cytokine tumor necrosis factor-alpha (TNF-α) have been effective

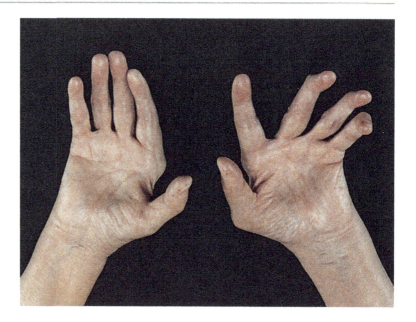

receptor, and its downstream consequences, will be analyzed in greater depth in Chap. 7 when interferon-β's interaction with interferon-α–β receptor (IFNAR) will be highlighted.

Cytokines are felt to be critical molecules in inflammation. Inflammation is usually defined by the physical findings on examination learned in Latin by every medical student: *rubor* (redness), *calor* (raised temperature), *tumor* (swelling), *dolor* (pain), and *functio laesa* (loss of function). Alternatively, inflammation is sometimes in addition defined by findings on microscopic examination of tissues, where the hallmark is infiltration of tissue by cells of the immune system such as the neutrophils, monocytes, and lymphocytes described above. Since joints are easily accessible to examination, it is relatively easy to measure the inflammation of arthritis, while inflammation within the nervous system is much harder to examine, and for the latter, the five cardinal signs described above do not apply in full. However, *functio laesa* (loss of function) is definitely one of the important findings of neuroinflammation, a point which will be further discussed in some of the neuroimmunological conditions described below. TNF-α has been implicated in inflammation (see above), but other cytokines are also important, including interferon-γ and interleukins 1, 6, and 8, and TGF-β.

The terms "inflammatory" and "immune-mediated" are sometimes used interchangeably, often because usually there is an increased presence of mononuclear cells such as lymphocytes and macrophages in both processes. However, they are not synonyms. For example, myasthenia gravis, which is clearly immune-mediated in that it is caused by autoantibodies to the nicotinic acetylcholine receptor (see Chap. 9), is not usually considered an inflammatory disease, since there is no redness, swelling, heat, pain, or influx of mononuclear cells. "Autoimmune" is also used loosely with the above terms, but this also should have a restricted meaning as discussed in greater depth in Chap. 3.

2.2.4   CD Antigens

CD antigens are surface molecules that are used to characterize immune cells. Sometimes, a given functional subset of immune cells can be identified by these molecules. For instance, cytotoxic T cells usually have the CD8 protein on their surface, so these cells are referred to as "CD8+" cells. Since all T cells have the CD3 marker, cytotoxic T cells can be separated from a wide variety of other cells by their combination of markers, CD3+CD8+. CD markers on cells are usually identified by monoclonal antibodies tagged with a fluorescent material identifying the

Table 1.3 CD markers

Cell category	CD marker
T cells	CD3
Helper T cells	CD4
Cytotoxic T cells	CD8
B cells	CD19, CD20
B1 cells	CD5
Memory B cells	CD27
Dendritic cells	CD8, CD11

CD marker of interest. After reaction of these labeled antibodies with the cells, the number of CD+ cells can be determined using a technique called fluorescent-activated cell sorting (FACS) in which the labeled cells are rapidly passed through a tube past a sensor which detects the fluorescence. This technology has been a great advance in characterizing immune cells and allowing their separation into various categories. A list of common immune cells and their CD markers is found in Table 1.3.

2.2.5 Cell Adhesion Molecules

When direct cell-to-cell interactions are essential, immune cells, as well as other types of cells, utilize a class of molecules called cell adhesion molecules (CAMs). CAMs are a large class of molecules which generally share the following characteristics. They have three domains: extracellular, transmembrane, and intracellular. The extracellular domain interacts with other cells, usually via CAMs on those other cells, or alternatively interacts with the extracellular matrix. In addition, the expression of CAMs is generally not constitutive, but inducible, often by cytokines which upregulate or downregulate CAMs. Although there are a number of structurally different subclasses within the CAMs, such as those called selectins , immunoglobulin superfamily, and cadherins, the class of CAMs most directly relevant to the clinical neuroimmunologist is in the subclass of CAMs called integrins. Alpha-4 integrin is the molecule targeted by natalizumab, a therapeutic monoclonal antibody utilized in MS therapy, which will be discussed in greater depth in Chap. 7. Alpha-4 integrin and other CAMs are essential in lymphocyte trafficking. CAMs are expressed by both lymphocytes and endothelial cell surfaces throughout the body, and the extent of expression of these molecules helps to regulate where mononuclear cells ingress and egress the circulation. Another group of molecules that affect lymphocyte trafficking are sphingosine-1-phosphate and its receptors explained in more detail in Chap. 19.

2.2.6 Complement

Complement is not a single molecule but a collection of over 25 proteins which work together to "complement" the ability of antibody to kill a pathogen. The initiation of the complement cascade occurs when an antibody binds to the pathogen. The end result is the activation of the membrane attack complex (MAC), which is able to damage membranes and results in the death of the pathogen. CD59, which is present on human cells, inhibits the MAC to prevent lysis of "self" cells, but some viruses, such as HTLV-1 and cytomegalovirus, make use of CD59 to prevent human-activated complement from lysing them. Another pathogen which utilizes complex pathways to avoid host killing by complement is *Borrelia burgdorferi*, the causative agent of human Lyme neuroborreliosis, which is able to express a variety of complement inhibitors, and upregulates complement inhibitors specific for the animal it is infecting [6].

References

1. Gershon RK, Kondo K. Infectious immunological tolerance. Immunology. 1971;21(6):903–14.
2. Gershon RK, Cohen P, Hencin R, Liebhaber SA. Suppressor T cells. J Immunol. 1972;108(3):586–90.
3. Cantor H. Reviving suppression? Nat Immunol. 2004;5(4):347–9.
4. Corthay A. How do regulatory T cells work? Scand J Immunol. 2009;70(4):326–36.
5. Dinarello CA. Immunological and inflammatory functions of the interleukin-1 family. Annu Rev Immunol. 2009;27:519–50.
6. Bykowski T, Woodman ME, Cooley AE, et al. Borrelia burgdorferi complement regulator-acquiring surface proteins (BbCRASPs): expression patterns during the mammal-tick infection cycle. Int J Med Microbiol. 2008;298 Suppl 1:249–56.
7. Hopkins DR. The greatest killer: smallpox in history. Chicago: University of Chicago Press; 2002. ISBN 0226351688.

Neurology for the Non-neurologist

2

The purpose of the nervous system is to interact with the environment to promote survival and reproduction. Thus, the nervous system must receive sensory inputs, the central nervous system (CNS) must "process" these inputs as well as internal signals and must respond via appropriate outputs many of them related to muscle movement. CNS "processing" includes such complicated phenomena as consciousness, emotion, and memory. The complexity of the nervous system is immense, a fact demonstrated by the expanding population of scientists involved in its study. The annual meeting of the Society of Neuroscience, basic scientists involved in the study of the nervous system, regularly attracts over 30,000 attendees. More clinically oriented neurology research meetings attract thousands of neurologists, physicians who care for patients with diseases of the nervous system. Since the focus in this chapter is neurological disease, rather than normal neurological function, there is not much basic neuroscience here, but the basis of all neurology is neuroscience, and the distinction between the two fields is somewhat artificial.

The human nervous system, as well as that of nonhuman primates, is an aberration among animals in its size and extravagant energy expenditure. One of the differences in the nervous system of primates relative to that of other animals is its dependence on vision instead of smell; the visual system requires a large amount of brain, but provides significant survival advantages to primates.

While accounting for only 2% of body weight, the brain utilizes approximately 20% of the total energy expenditure.

1 Organization of the Nervous System

The nervous system is generally divided into the CNS, consisting of the brain, brain stem, cerebellum, and spinal cord, and the peripheral nervous system (PNS), consisting of the nerve roots, plexi, peripheral nerves, the neuromuscular junction, and the muscle. The flow of sensory information is from the periphery toward the brain, and the motor output is from the brain peripherally. This "brain outward" flow of motor output in the nervous system is also called "rostrocaudal" flow. A small minority of sensory inputs with motor responses, requiring great speed, can bypass the brain and are called reflexes.

1.1 Electrical Nature: Nerve Transmission and Neurotransmitters

The nervous system is electrical, and information is composed of electrical signals which are transmitted along nerve cells or across synapses. The first demonstration that the nervous system was electrical was by Luigi Galvani, an Italian physician,

A.R. Pachner, *A Primer of Neuroimmunological Disease*,
DOI 10.1007/978-1-4614-2188-7_2, © Springer Science+Business Media, LLC 2012

in 1771 who demonstrated that an electrical impulse applied to a frog's leg made it kick. The electrical signal which passes along a nerve is called an action potential or nerve impulse and is a very fast, transient change in voltage across the nerve membrane mediated by specialized molecules in the membrane, voltage-gated ion channels. The speed of movement of the action potential averages about 50 m/s in the human or 110 miles/h. The membrane recovers very quickly and can generate many impulses per second. These electrical signals are transmitted by specialized cells called neurons. Communication between neurons is accomplished by synapses in which the arrival of an action potential results in the release of neurotransmitters, specialized molecules that can be quickly moved and processed, across the synapses leading to an action potential in the downstream neuron. The nervous system uses many different types of synapses which can be differentiated by location within the nervous system as well as by the type of neurotransmitter. There are many different types of neurons in the CNS, defined by their shape and their variety of synapses. One particularly striking and important type of neuron is the Purkinje cell, first described in 1837 by the Czech anatomist, Jan Purkinje, and localized to the cerebellum. This cell is a target in the disease paraneoplastic cerebellar degeneration (see Chap. 14).

The human nervous system utilizes a wide variety of neurotransmitters, which for any synapse can be considered excitatory or inhibitory in their effect on transmission of the impulse. The most well understood is acetylcholine, mentioned below, which is used both in the central and PNS. The most prevalent neurotransmitter is glutamate, which is usually excitatory, while gamma-amino butyric acid (GABA) is usually inhibitory. GABA is, in addition, an important neurotransmitter for the neuroimmunologist because GABA agonists, such as baclofen, are used in the treatment of spasticity caused by the most common neuroimmunological disease, multiple sclerosis (MS) (see Chap. 7). Monoamines, such as dopamine, epinephrine, and norepinephrine, are also important neurotransmitters; the adrenergic agonist tizanidine is also used in MS.

1.2   Cells of the Nervous System

A typical neuron possesses a cell body (often called soma), dendrites, and an axon (see Fig. 2.1), arranged in a variety of configurations. The axon is the main electrical cable and is highly specialized for action potential transmission, while the cell body provides the housekeeping functions for the whole cell, and the dendrites communicate with other neurons via synapses. Axons can be very long; in the human, axons traveling in the sciatic nerve, which have their cell bodies in the spine and extend to the feet, can be four feet in length, while many nerve cells in giraffes are much longer. Proteins from the cell body to distal parts of the axon are transported along axonal neurofilaments via a class of molecules called kinesins, considered to be "molecular motors." The axon of most neurons, but not the cell body, is covered by myelin, which provides insulation. This allows the electrical signal to be saltatory, or hopping, between specialized unmyelinated areas along the

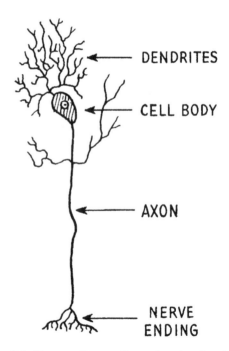

Fig. 2.1 Neurons. Nerve cells conduct impulses in a directed manner, from the dendrites through the axon to the nerve ending. The position of axons and dendrites relative to the cell body and to each other may vary (Wikimedia, public domain, at http://commons.wikimedia. org/wiki/File:Dendrite_%28PSF%29.png)

axon called nodes of Ranvier, which are present at intervals along the axon, discovered in 1878 by Louis-Antoine Ranvier. Myelin is mostly lipid in content, while the cell body is mostly proteinaceous, leading to a distinctive whiteness in CNS areas where myelinated axons are preponderant, and to grayness where cell bodies mostly occur. This distinction leads to the commonly used terms for the gross appearance of areas within the brain and spinal cord, "white matter," where there are myelinated axons, or "gray matter," for areas where there are mostly neurons. Myelin is not a component of the neurons themselves, but of specialized myelin-producing cells. In the CNS, myelin is produced by oligodendrocytes ("cells with few dendrites") and in the PNS, it is produced by Schwann cells first described by Theodor Schwann in the nineteenth century. Myelin, a complex mixture of lipid and proteins, is different biochemically in the CNS versus the PNS. Myelin is essential for efficient transmission of impulses, similar to the way insulation around a wire is essential to efficient transmission of electricity along the wire. Demyelination, the loss of myelin during diseases is a major feature of two common neuroimmunological diseases, multiple sclerosis (Chaps. 4–8) and Guillain–Barré syndrome (Chap. 9). Most of the protein in central myelin is proteolipid protein (PLP) which is abnormal in the genetic disease, Pelizaeus–Merzbacher disease (see Inset 2.1), while in peripheral myelin the major protein is myelin protein zero (MPZ). Both central and peripheral myelins contain myelin basic proteins (MBPs) and myelin-associated glycoprotein (MAG), two proteins which will be discussed in later chapters.

Inset 2.1 Pelizaeus–Merzbacher Disease: A Genetic Disease of Myelin [4]

At-risk population, cause, symptoms/signs. The disease as originally described in Germany in the late nineteenth century is a genetic X-linked disease, beginning in infancy, resulting in slow, progressive loss of motor control, ataxia, spasticity, and cognitive impairment, characterized pathologically by the loss of myelin and oligodendrocytes. The cause is a problem with the gene for PLP; these abnormalities range from missense mutations to the more common duplications of the entire *PLP* gene, which account for the majority of cases of Pelizaeus–Merbacher disease. The absence of a clear understanding of the role of PLP in healthy myelin function, and the wide variety of genetic alterations in this gene, have led to great difficulty in genotype–phenotype correlations.

Morbidity/mortality. The disease is fatal in its most severe form. However, many patients with abnormal PLP genetics can have relatively mild disease. One form is the slow development of spastic paraplegia, called spastic paraplegia 2 (SPG2), which can be quite mild.

Treatment. There is no effective treatment. Although gene replacement may help some patients, it will not help most, since the most common cause is gene duplication.

Prevention. Genetic counseling should be made available to the families of patients with PMD or SPG2.

Oligodendrocytes are members of a family of cells in the CNS called glia, Greek for "glue"; other glia are astrocytes and microglia. There are approximately equal numbers of glia and neurons in the human CNS, with the most common glial cell being the astrocyte. The microglia of the brain can be considered the brain macrophages and constitute about 20% of all brain cells. Microglia are derived from hematopoietic precursors, can multiply in response to an antigen stimulus, and are mobile within the CNS. They are extremely changeable, being able to accommodate quickly to changes in the microenvironment within the CNS. They can assume multiple morphologies depending on their level of activation. Astrocytes are not motile and have diverse functions, thought to be mostly in support of neurons. They are able to communicate with each other using calcium waves through gap junctions; the role of this communication is unknown [1]. Astrocytes are the cells thought to be destroyed first in the neuroimmunological disease neuromyelitis optica (NMO) described in Chap. 6.

1.3 Structure of the Nervous System: CNS, PNS, Upper and Lower Motor Neurons

The CNS is divided structurally into three areas which have different functions: the brain, brainstem, and spinal cord (Fig. 2.2). The brain, consisting of the cerebrum and cerebellum, is the central processing center for sensory input and motor output, as well as the source of consciousness, memory, emotion, and thought. Just below the brain is the brainstem, the processing center for the cranial nerves, which account for critical functions related to vision, hearing, facial movement sensation, swallowing, and speech. The spinal cord is the local processing center for the

functions of the arms and legs, as well as bowel and bladder function. The cerebrum is divided into specific lobes, frontal, temporal, parietal, and occipital lobes, each of which becomes specialized for certain brain functions, as the brain develops both in utero and after birth. The primary motor cortex, activated for the purpose of muscle movement, is in the frontal lobe, while the primary sensory cortex, the initial site of cortical processing of sensory information, is located in the parietal lobe. Language function is located in the temporal lobe, while the very large occipital lobes of the human are devoted to vision and its processing. Most motor fibers cross sides within the lower part of the brainstem, the medullary decussation, while sensory fibers cross within the spinal cord. Thus, a destructive lesion in the right parietal lobe will result in sensory loss on the left side of the body, while damage to the left frontal lobe, will result in weakness of the right side of the body. The speech centers of the brain are located in the left temporal lobe, and destructive lesions there will cause aphasia, loss of language function. The brain consists of two broadly different types of neurons, those involved in communication within the brain, called interneurons, which are by far the most common, and those involved in communication with caudal structures, called projection neurons.

The CNS is protected from trauma by bone. The brain and brainstem is surrounded by the skull, and the spinal cord by the vertebrae. The vertebral column, also called the spine, consists of 24 articulating vertebrae divided in sections called cervical, thoracic, and lumbar spine. There are nine fused vertebrae below the lumbar vertebrae in the sacrum and coccyx. To provide mobility to the spine, the vertebrae are connected to each other posteriorly through specialized joints and anteriorly are stacked one above the other, with cushions in between called disks. Sometimes disks can become damaged, herniate, and compress nerve roots or spinal cords (see Inset 6.1).

The PNS consists of all parts of the nervous system peripheral to the CNS and includes nerve roots, plexi of roots and nerves (especially the brachial and lumbar plexus), peripheral nerves, neuromuscular junctions, and muscle. Since each muscle is

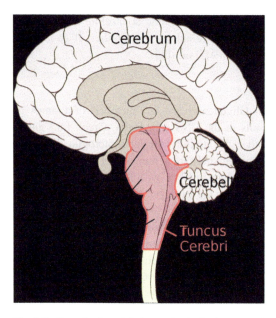

Fig. 2.2 Organization of the human brain, brainstem, and spinal cord in sagittal section, with brain stem highlighted (from Patrick Lynch http://commons.wikimedia.org/wiki/File:Brain_sagittal_section_stem_highlighted.svg)

innervated by one nerve, injury to nerves can be detected by electrical testing of muscles, called electromyography (see Inset 2.2). The sensation of touch is also transmitted through nerves according to specific areas assigned to specific nerves, e.g., two major nerves lead to the hand, the median and the ulnar, both of which are relatively superficial and susceptible to injury (Inset 2.3). Transmission of signals from motor nerves to muscle is effected by a specialized synapse called the neuromuscular junction, in which the neurotransmitter is acetylcholine, and the receptor on muscle which captures and processes released acetylcholine, is the acetylcholine receptor (AChR). The AChR is a target of autoantibodies in the neuroimmunological disease myasthenia gravis (see Chap. 10).

progressive nature and the large number of individuals affected.

Treatment. Generally ineffective. Various pain medications, including anticonvulsant type drugs, are used to give some symptomatic relief.

Prevention. It is generally believed, but not proven, that careful control of blood sugar can delay or prevent diabetic neuropathy.

Inset 2.3 Clinical Problem—The Hand: The Funny Bone and Carpal Tunnel Syndrome

---

At-risk population. The human hand is a wonder of biology, but the nerves innervating the hand are prone to injury.

Cause. The most common neuropathy is a transient one, caused by "hitting the funny bone," when the ulnar nerve is traumatized, as it travels through the elbow and passes very close to the skin at the medial epicondyle of the humerus. Chronic trauma can lead to permanent injury. Another common neuropathy is carpal tunnel syndrome, or CTS, in which the median nerve is traumatized by constant pressure against the tendons in the wrist.

Symptoms and signs. These two conditions lead to numbness and tingling in the hand, the particular area of which depends on the innervation associated with the nerve.

Morbidity/mortality. Each of these conditions, which are very common, can cause dysfunction of the hand and are common causes of disability in our society, where working with hands, especially keyboards, has become so important.

Treatment. Surgery can result in the protection of the nerve from future injury.

Prevention. Repetitive trauma to the arm can be hard to prevent but some simple measures such as using cushioned arm rests may be helpful.

Inset 2.2 Clinical Problem—Diabetic Neuropathy and the Modern Equivalent of the Galvani Experiment

---

At-risk population. Approximately 3% of the world's population suffers from diabetes, a large percentage of which has neuropathy, i.e., nerve damage causes numbness, weakness, and pain.

Cause. Diabetes mellitus is a disorder of insulin production or response, and, by mechanisms that are not understood, is associated with injury to neurons in the PNS but not in the CNS.

Symptoms and signs. Patients with diabetic neuropathy typically have symptoms in a "stocking-glove" distribution, usually consisting of pain, numbness, and tingling. As the disease worsens strength in the legs and then arms decreases, and walking can be impaired.

Patients with diabetic neuropathy can be diagnosed using electrical stimulation of tissue by electromyography/nerve conduction (EMG/NCV) studies, a commonly used neurological test which is a modern modification of Galvani's stimulation of the frog leg.

Morbidity/mortality. The burden to society of this disease is substantial, because of its

Table 2.1 Clinical differences between UMN and LMN lesions

Characteristic	UMN	LMN
Location of injury	CNS	PNS (except for anterior horn cell in spinal cord)
Strength	Decreased	Decreased
Muscle tone	Increased	Decreased
Muscle bulk	Normal	Decreased
Deep tendon reflexes	Increased	Decreased
Sensory abnormalities	Variable	Variable

The motor system of the body can also be conceptually divided into upper motor neuron (UMN) and lower motor neuron (LMN) territories. UMNs are always in the CNS and project onto LMNs in the brain stem spinal cord. The anterior horns of the spinal cord contain the cell bodies of these LMNs which send axons that innervate muscles. Most diseases selectively target one location in the CNS and frequently have distinctive UMN or LMN clinical presentations (Table 2.1). One disease that has a combination of UMN and LMN findings is amyotrophic lateral sclerosis (ALS) which damages anterior horn cells and also UMN pathways within the spinal cord (see Inset 2.5). Stroke and multiple sclerosis which are CNS disorders manifest only UMN findings.

The dominant sensory input in humans is vision, a characteristic that distinguishes primates from most other animals. A large part of the human CNS is devoted to vision and its processing. The optic nerve, which carries signals from the retina to the brain, is actually not really a true nerve, but an extension of the CNS. The optic nerves, i.e., cranial nerves II, lead to the occipital lobe where visual images are processed; the occipital lobe is the only part of the brain devoted entirely to one function, vision. Inflammation of the optic nerve is common in neuroimmununological diseases, especially multiple sclerosis (see Chap. 4). Optic nerve inflammation is called optic neuritis and results in the loss of vision and eye pain. Smell is brought to the brain from the nose via cranial nerve I, the olfactory nerve; touch on the face via cranial

nerve V; and hearing via cranial nerve VIII. Touch, position, and vibration sense in the body come to the brain from peripheral nerves to the spinal cord and then via the spinothalamic tract to the brain. The sensory inputs come from specialized receptors called mechanoreceptors for touch and vibration, including Meissner's corpuscles, Pacinian corpuscles, and Merkel disk receptors, and nociceptors for pain. Position sense, identifying position in space with or without movement, is mediated by a group of specialized receptors in muscles, tendons, and joints.

The CNS is surrounded by a number of membranes. The one immediately overlying the CNS parenchyma is the thin and delicate pia mater, which accompanies blood vessels as they penetrate into the brain; the pia mater is a part of the neurovascular unit [2]. The neurovascular unit is a structure composed of components of both blood vessels and CNS cells that determines, among other things, the blood–brain barrier (BBB) and is further described in Chap. 3. The BBB and the blood–CSF barrier consist of cellular barriers between blood and CNS parenchyma that prevent the easy flow of large or highly charged molecules between the blood and the CNS. Between the pia and the next layer, the arachnoid mater, is a space called the subarachnoid space. Over the arachnoid is the thick dura mater, and the space under the dura is thus called the subdural space. The spaces defined by these membranes are used frequently in describing pathological problems, especially bleeding. Bleeding within the brain can either be localized to the brain parenchyma and not spread elsewhere, in which case it is considered an intracerebral hemorrhage (ICH), or it can be restricted to the subarachnoid space [subarachnoid hemorrhage (SAH)] or subdural space [subdural hemorrhage (SDH)].

The brain is suspended in a fluid, called cerebrospinal fluid (CSF). After being produced by ependymal cells, epithelial cells that line the ventricles, this clear, colorless fluid, circulates around the CNS in the subarachnoid space, i.e., between the arachnoid and pia maters, and eventually becomes resorbed by the arachnoid villi or drained into lymphatics. In many diseases of

Table 2.2 Some examples of CSF abnormalities in disease processes

Disease	CSF abnormality
Inflammation (e.g., encephalitis)	Increased number of white blood cells
Infection (e.g., pneumoccal meningitis)	Presence of the pathogen
Cancer (e.g., lymphoma)	Presence of cancer cells
Bleeding in the brain (e.g., burst aneurysm)	Presence of blood
Increased pressure in the brain (e.g., hydrocephalus)	Increased pressure

the nervous system, there are characteristic abnormalities of the CSF (see Table 2.2), and the physician can easily sample CSF by performing a lumbar puncture in which a needle is inserted into the subarachnoid space.

2 The Neurological Evaluation

2.1 Neurological Examination

When disease results in neurological symptoms, neurologists are frequently called to evaluate the patient, identify the diagnosis, and recommend appropriate therapy. The initial evaluation consists of obtaining a history, and performing a neurological examination. Diagnoses entertained after this evaluation are further pursued, if necessary, by laboratory testing, including blood testing, imaging, and other tests.

Neurological symptoms usually consist of difficulty with some aspect of nervous system function that we normally take for granted. The most common are problems with ambulation, coordination, strength, and the sensory functions of vision or touch. Humans are normally awake, alert, and can perform difficult cognitive tasks; neurological disease can impair level of alertness and interfere with cognition.

The first part of the neurological examination involves assessing higher cognitive functions, sometimes utilizing the "Mini-mental Status" examination [3]. Patients who have diffuse dysfunction of gray matter in the brain affecting concentration, attention, memory, recall, etc. are said to have an encephalopathy. Next, the cranial nerves (CNs) are tested: first, the optic nerve, CN II, is tested by the examiner using an ophthalmoscope, a small hand-held instrument, looking at the optic nerve head on fundoscopic examination, then by assessing vision. Eye movements are then checked, usually by having the patient follow the examiner's moving finger. Abnormalities in the conjugate movement of the eyes are common and result in the symptom of double vision, also called diplopia; sometimes diplopia is not present, but instead of smooth pursuit of a moving finger, the eye movements occur in jerky, oscillating movements called nystagmus. Eye movement testing, and testing of the pupillary response to light, tests CNs III, IV, and VI. Normal eye movements are conjugate, meaning each eye is moved equivalently. Conjugate eye movement can be impaired either by a lesion affecting one cranial nerve or by lesions of the brainstem where there are extensive connections among cranial nerves to allow for conjugate eye movements. The medial longitudinal fasciculus (MLF) is one of the connections in the brain stem and is often injured in multiple sclerosis (see Chap. 4). CN V is assessed by determining the presence of normal facial sensation and under some circumstances normal corneal reflex, the rapid shutting of the eyelid when the cornea is touched. Movement of the face tests CN VII, and hearing tests CN VIII. CN X is tested by determining that the palate moves normally and the voice is normal. CN XI is tested by measuring the strength of the sternocleidomastoid and trapezius muscles, two large muscle groups of the neck, while CN XII controls tongue movement. The motor system is tested by evaluating the strength of proximal and distal muscles in the four extremities, and by testing muscle tone and bulk. Coordination is tested by analysis of the ability to perform well-controlled movements such as rapidly moving the finger from the examiner's finger to the patient's nose. The sensory system is then examined using tests of light touch, vibration, position sense, and pin prick. Deep tendon reflexes are then tested in the arms and legs, followed by looking for any abnormal reflexes such as the

Babinski or Hoffman. The Babinski reflex is a time-honored bedside test in neurology, initially developed by Joseph Babinski, a late nineteenth century French neurologist, to differentiate psychiatric from neurological disease (Inset 2.4). Injury to descending tracts in the brain or spinal cord result in increased deep tendon reflexes and increased muscle tone. Gait is then tested, including walking one foot right before the other ("tandem gait"); inability to perform tandem gait because of problems with balance is called ataxia, a common problem in multiple sclerosis. The inability to walk well because of weakness or ataxia or abnormal muscle tone is a characteristic of many neuroimmunological diseases and is the function most important in determining the extended disability status score (EDSS), a disability scale used for measuring multiple sclerosis (see Chap. 4).

Inset 2.4 Differentiating Between "Structural" and "Psychiatric" Disease—The Babinski Reflex

Bizarre behavior is frequently caused by psychiatric disease, especially schizophrenia, which has a lifetime prevalence in the population of between 0.5 and 1%. However, behavior changes can also be called by "structural" neurological diseases, such as tumors and infections of the brain. Neurologists often struggle with making the proper diagnosis. They frequently use a test described in 1896 by Joseph Babinski, a French neurologist of Polish origin, who practiced with Charcot at the Saltpetriere in Paris. The Babinski test involves stroking the sole of the foot, and eliciting a movement of the toe upwards in individuals with damage to the corticospinal tract, but downward movement in the absence of such injury. A positive Babinski, i.e., an extensor upward movement of the toe, can be a very early sign of injury, but will be absent, i.e., a flexor downward movement, in patients with psychiatric disease.

The combination of the history and neurological examination will allow the neurologist to begin answering the two most critical questions when faced with a patient with neurological symptoms: where and what? Where is the lesion? And, what is the most likely process causing dysfunction there? For instance, the sudden onset of a left body weakness and sensory loss in a 70-year-old man with a long history of vascular disease, hypertension, smoking, and diabetes would most likely be a right cerebral stroke, while the location would be the same for a 30-year-old woman with no medical history, but the lesion causing it might be more likely to be multiple sclerosis rather than a stroke.

2.2 Imaging of the Nervous System

The neurologist frequently will confirm the likely diagnosis by some further testing. The standard tool for imaging the nervous system is computerized tomography (CT) scanning, first available for patients in 1972, which involves computer analysis of X-ray signals. Almost all hospitals in the USA have CT scanners. Another method of imaging is magnetic resonance imaging (MRI), which does not use X-rays, but instead magnetization as the primary source of signal. The principle for most types of MRI scans is that the water molecule is a dipole with the hydrogen nuclei producing a magnetic field after being aligned by an induced magnetic field or radiofrequency signals. The combination of the magnetic fields, the radiofrequency signals, and the timing of the scanning creates many potential ways to image living tissues. MRI is generally more sensitive and specific than CT scanning and does not have the potentially damaging effect of radiation dose caused by X-rays. However, it is much more expensive than CT scanning, and is not as widely available. The MRI has been especially useful in multiple sclerosis; its use in that disease will be discussed extensively in later chapters. A type of MRI called functional MRI may also become clinically useful in assessing recovery after CNS injury (see Inset 2.5). CT and MRI scan not only show areas of abnormality by their characteristic signals but also show the absence of CNS tissue; many disease processes, especially those with a degenerative component such as Alzheimer's disease and multiple sclerosis, lead to the loss of CNS tissue which is demonstrable on imaging as atrophy.

Inset 2.5 Clinical Problem—Recovery After CNS Injury, Neural Plasticity, the Contribution of Kittens and Phantoms, and Functional MRIs

---

CNS injury is unfortunately a common problem in the society, with motor vehicle accidents and strokes common causes. Neurons do not divide, and dead neurons generally cannot be replaced with new neurons. Thus, recovery of function after injury must involve processes other than growth of new cells. One of these mechanisms is neural plasticity, in which neurons involved in other functions change to pick up the lost functions from injury. The concept of plasticity and other processes are obviously of considerable interest to clinical neuroscientists who wish to optimize recovery after injury, but our knowledge of these processes remains relatively primitive.

It was thought that brain regions were "hard-wired," but this was proven wrong by pioneering experiments in the 1960s and 1970s by Hubel and Wiesel [5]. In kittens who from birth had one eyelid sewn shut, the primary visual cortex receiving inputs from the functioning eye took over the areas that normally received input from the deprived eye.

In individuals who have an amputation, rubbing of the face or lips can induce sensations perceived as if they were in the amputated, i.e., absent, limb. This is called phantom sensation, and when it is associated with pain, phantom pain. This appears to be because of downward shift of the hand area of the sensory homunculus onto the area of face representation, especially the lips, sometimes called functional cortical remapping.

The field of cortical remapping and determining whether humans can build new structural connections is an active and in 2011, a very controversial area. One of the tools of investigators in this field is functional MRI, in which activation of brain regions can be imaged by the increased blood flow caused by increased neuronal activity.

Some atrophy is due not primarily to death at the site of the atrophy visualized on imaging, but may be due to injury to a part of the nerve cell distant from the atrophy. Wallerian degeneration is axonal atrophy below an area of injury to the neuron. If the axonal injury is healed the healthy axon above the injury can grow back into the intact neurolemma, the hollow myelinated tube of the nerve, which is part of the Schwann cell or oligodendrocyte, and may not be affected by the injury. Sometimes complete recovery can occur if Wallerian degeneration goes smoothly and the neurolemma remains intact and healthy. This happens more commonly in PNS than CNS diseases.

2.3 EEG and EMG

Many patients with brain disease develop seizures or impairment of the electric signals in their brains detectable by the measurement of electric field potentials measured over the scalp, known as electroencephalography (EEG). Since measurement is over the scalp and electric fields decay rapidly with distance, the EEG mostly assesses cortical neurons close to the skull. Most of the field frequencies are about 3–20 Hz, and the amplitudes are usually 10–80 μV. The relatively low voltage is due to a great extent from damping from the skull since the signals from subdural electrodes are nearly three logs higher in amplitude than those from surface electrodes. Characteristic patterns are seen in normal awake, drowsy, and sleeping adults, and the patterns change depending on age of the patient, and location on the skull of the electrodes tested. The most dramatic abnormalities are seen during seizures when synchronous bursts of activity can be seen as spike and wave forms which replace the normal rhythms. EEG-detected seizures may or may not have associated clinically obvious sequelae.

The electrical activity of the nervous system can also be used to detect abnormalities in the peripheral system using a test known as electromyography and nerve conduction velocity testing or EMG/NCV. In EMG, needles are inserted into muscles thought to possibly be involved in the pathological process. In a patient with ALS, Lou Gehrig's disease (see Inset 2.6) insertion of a

Inset 2.6 Amyotrophic Lateral Sclerosis—UMN and LMN Findings

At-risk population. An uncommon disease affecting individuals of both sexes usually between 30 and 50 years old. The annual incidence is about 1 in 100,000.

Cause. Unknown.

Symptoms and signs. Progressive muscle weakness and atrophy. The disease is often called Lou Gehrig's disease. One of the best baseball players ever, Lou Gehrig held many records for both batting (most grand slams, most homers in a game, etc.) and for durability (most consecutive games played), until May 2, 1939, when he benched himself and sought medical attention because of his progressive weakness and was diagnosed with ALS. His retirement speech at Yankee Stadium on July 4, 1939 ("I consider myself the luckiest man on the face of the earth") was one of the most eloquent by a sports figure. He died of ALS 2 years later.

Morbidity/mortality. The disease is fatal, usually leading to death in 2–5 years after diagnosis.

Treatment. No treatment has been shown to be effective and there is no way to prevent the illness.

recording needle into weak muscles will demonstrate characteristic abnormalities referable to the involvement of these LMNs by the disease in the anterior horn cells of the spinal cord. NCV testing determines the velocity of nerve conduction which for most nerves is approximately 50 m/s. The most dramatic slowing of nerve conduction velocities is seen in conditions which result in the loss of myelin, such as acute or chronic inflammatory demyelinating polyneuropathy (AIDP or CIDP), conditions discussed in Chap. 9. In contrast, processes which injure axons in the PNS but generally spare myelin, e.g., the neuropathy of diabetes mellitus, will result in lower and deformed action potentials out of proportion to slowed nerve conductions.

References

1. Giaume C, Koulakoff A, Roux L, Holcman D, Rouach N. Astroglial networks: a step further in neuroglial and gliovascular interactions. Nat Rev Neurosci. 2010; 11(2):87–99.
2. Owens T, Bechmann I, Engelhardt B. Perivascular spaces and the two steps to neuroinflammation. J Neuropathol Exp Neurol. 2008;67(12):1113–21.
3. Folstein MF, Folstein SE, McHugh PR. Mini-mental state. A practical method for grading the cognitive state of patients for the clinician. J Psychiatr Res. 1975;12(3):189–98.
4. Garbern JY. Pelizaeus-Merzbacher disease: genetic and cellular pathogenesis. Cell Mol Life Sci. 2007; 64(1):50–65.
5. Hubel DH. Exploration of the primary visual cortex, 1955-78. Nature. 1982;299(5883):515–24.

Neuroimmunology for the Non-neuroimmunologist

3

After two exceedingly brief reviews of immunology and neurology, we are ready for an exceedingly brief review of neuroimmunology. Neuroimmunology is a nascent field and our understanding of most of the field is relatively shallow. This chapter serves as a brief historical review of trends in the field and an introduction to the remainder of the book.

Neuroimmunology is focused on disease, since the nervous system is not thought of as participating in immune functions under "normal" conditions. There is no neuroimmunological system development, unlike its parent systems each of which has their own developing structures, cells, and locations. Thus, part of the title of this book, "*Neuroimmunological Disease*," can be considered redundant, since neuroimmunology is usually thought of in the context of disease. But this is not surprising for an infant field; immunology and neuroscience were both born out of human disease, in attempts to understand diseases such as syphilis and dementias. At some point in the future, the nervous system will be considered an essential functioning part of the normal immune system, and the immune system will be considered an essential part of the normal nervous system. These overlaps are already being studied. The autonomic nervous system innervates lymphoid organs, but how this functions in normal functioning of the immune system is unknown. Conversely, a classical "immune" cell, the microglial cell, is a normal constituent of the central nervous system, in fact constituting about 20% of the total number of glial cells, but its function in normal CNS function is essentially

unknown, but is an area of intense study. Thus, at this point in time, neuroimmunology is generally a study of disease states, and we are hopeful that our knowledge will expand into understanding normal functions as well.

Given the difficulty of accessibility of human nervous system tissue, the study of neuroimmunology has historically been primarily the study of animal and cellular models of disease. This focus on nonhuman systems has its advantages and disadvantages. An advantage is that information can be much more rapidly accrued about the interactions of the immune and nervous systems. A disadvantage is that the human nervous system is very different from the nervous system of lower animals, with the possible exception of nonhuman primates. Thus, the relevance to humans of knowledge gained from experiments on neuroimmunology in mice, neuroimmunology's favorite animal for models, is not as high as it might be for other organ systems in which there is more of a resemblance to humans.

1 The Beginnings of Neuroimmunology: Post-vaccinial Encephalomyelitis

The field of neuroimmunology had its conceptual beginning in the late nineteenth century during work on rabies, a fatal neurological infection (see Chap. 12 for more on rabies), when it was found that some individuals developed severe

A.R. Pachner, *A Primer of Neuroimmunological Disease*,
DOI 10.1007/978-1-4614-2188-7_3, © Springer Science+Business Media, LLC 2012

Table 3.1 Experimental models of presumed neuroimmunological disease induced by immunization of animals with nervous system components

Immunization with:	Name of model	Passive transfer/ active immunization	Purified antigens used	Cellular or humoral	Animal	Disease modeled
CNS	EAE	Both	MBP, PLP, MOG, and their peptides	Cellular	Multiple	ADEM
PNS	EAN	Both	P0, P2, SGPG, and their peptides	Cellular	Rodents	GBS, CIDP
Acetylcholine receptor (AChR)	EAMG	Both	AChR	Humoral	Multiple	Myasthenia gravis

CNS inflammation after the rabies vaccine was administered, i.e., the immune system could damage as well as protect. Although rabies incidence has decreased substantially in the early twenty-first century with the availability of very good vaccines, there are still today about 50,000 deaths from rabies worldwide reported per year, and many rural and suburban areas in the USA have persistent reservoirs of rabid wild animals. When Louis Pasteur and Emile Roux developed the rabies vaccine in 1885, they used desiccated spinal cords of infected rabbits as the vaccine material. Later rabies vaccines also used CNS tissue; the Semple rabies vaccine is an inactivated homogenate of the brain of goats or sheep infected with rabies virus. The Semple vaccine, or other vaccines purified from animal CNS, which contain CNS material, continues to be used in some countries, including India, and post-vaccinial encephalomyelitis continues to be a clinical problem occurring in about one in 220 individuals receiving the Semple vaccine [1] usually 2–3 weeks after the vaccine. Post-vaccinial encephalomyelitis, in this case Semple rabies vaccine-associated encephalomyelitis (SAE), is a subset of a group of diseases in humans called acute disseminated encephalomyelitis (ADEM). The realization that immunization with CNS tissue could induce an inflammatory encephalomyelitis in humans is the basis for a group of neuroinflammatory models in experimental animals, collectively called EAE (experimental autoimmune encephalomyelitis), which have accounted for a large percentage of the research effort in neuroimmunology. ADEM and EAE will be extensively discussed in later chapters in this book.

The concept that antigens within the nervous system could be purified and used as immunogens for experimental models has served as a guiding light for the development of multiple animal models in neuroimmunology [2]. In the past 50 years, as immunology made major advances, the number of models multiplied; the major models have been EAE, experimental autoimmune neuritis (EAN), and experimental autoimmune myasthenia gravis (EAMG) (see Table 3.1). A variety of other models have been developed, none of which have been used as extensively as those listed in the table. A partial list includes: models of neuromyelitis optica (NMO), limbic encephalitis, Lambert–Eaton myasthenic syndrome (LEMS), Rasmussen's encephalitis, stiff-person syndrome (SPS), and pediatric autoimmune neuropsychiatric disorders associated with streptococcal infection (PANDAS).

Historically, the pathway for new ideas and new potential treatments in neuroimmunology has been from bench-to-bedside, but over the last decade there has been a great deal of new knowledge gained from testing potential targets in human disease, usually using monoclonal antibodies that have been developed for other diseases, and then going back to animal models and cellular systems when the new therapies have demonstrated therapeutic efficacy [3]. One of the best examples of this is the B-cell depleting monoclonal antibody, Rituximab, which demonstrated considerable efficacy in treating MS, and has led to a flurry of research on B cells in MS; MS had previously been considered by many to be exclusively a "T cell-mediated disease." This "bedside-to-bench" flow of information will be discussed at greater length when monoclonal antibody therapy is discussed. Another area in which "bedside-to-bench" flow of information has occurred has been in the increasingly broad field of autoantibodies. Academic neuroimmunology clinical laboratories, which had previously been primarily used for the detection of autoantibodies in paraneoplastic syndromes (a topic to be reviewed in Chap. 14), have extended their interest to non-paraneoplastic diseases. The most dramatic discovery has been the finding this past decade that NMO, thought to be a rare variant of multiple sclerosis, is actually a disease mediated by anti-aquaporin 4 autoantibodies with diagnostic and therapeutic features distinct from MS. This "bedside" observation has brought "bench" researchers to focus on aquaporin channels as biologically relevant molecules.

The animal model that dominates neuroimmunology by sheer numbers is EAE. As of 2010, the number of articles referenced on PubMed for experimental allergic encephalomyelitis (EAE, also called experimental autoimmune encephalomyelitis) is 3,554, far outnumbering EAN (219), and EAMG (461). A major concept in neuroimmunology critical for the understanding of EAE is that of communication between the CNS and the periphery related to the vasculature. The CNS is thought to have a relatively primitive innate immune response available from its own cells, and thus must "import" cells from the periphery when necessary. In EAE, immune cells are activated in the peripheral lymphoid tissues and then enter the CNS in increasing numbers. The area of the CNS in which these peripherally activated immune cells enter the CNS in EAE and other neuroimmunological processes is called the neurovascular unit. The larger blood vessels (#6 in Fig. 3.1), located in the subarachnoid space (#7), enter the parenchyma of the nervous system from the outside as penetrating vessels (#11). The cell body, also called soma or perikaryon, of astrocytes near the surface of the brain (#1), send out processes which terminate in endfeet; endfeet of many astrocytes near the blood vessel, and the subarachnoid space are joined in a membrane called the glia limitans (#2), which in this location is called the glia limitans superficialis. Endfeet connecting to each other near blood vessels deeper within the parenchyma (#13) are

called glia limitans perivascularis. The perivascular space, "PVS" in Fig. 3.2, is bounded by the glia limitans and by the walls of post-capillary venules. In capillaries of deeper sections of the brain (#12), shown by the black bar, the perivascular space vanishes, and the capillary endothelium ("E"), fuses with a specialized cell called the pericyte ("Pe") and then the astrocytic endfeet (white arrows in Fig. 3.2) to create a "fused gliovascular membrane." Aquaporin 4, the putative autoantigen of NMO (see Chap. 6), is found in the astrocytic endfeet.

Normally, in the absence of infection or inflammation in the CNS, there is trafficking of a small number of immune cells through the CNS. Under inflammatory conditions, trafficking increases and cells which have receptors specific for new antigens present in the CNS, either specific for pathogens in infections or CNS antigens in EAE, will linger in the CNS and become activated. As more cells become activated, cytokines are upregulated, as well as a specialized class of cytokines, called chemokines which are felt to be key chemoattractant molecules for immune cells. The combination of the presence of these chemokines as well as a transiently breached blood–brain barrier continue to recruit more cells. Entry of cells into the CNS appear to be via a two-step process. Initially, the cells in the earlier phases of neuro-inflammation accumulate in the perivascular space, with few immune cells from the blood entering into the parenchyma. However, as the process proceeds, especially if it is poorly controlled, or unusually strong, more and more inflammatory cells enter into and remain in the CNS parenchyma.

Our understanding of the immune response within the CNS and PNS remains primitive. Compared to the power of immune response outside of the CNS, the immune response within the CNS appears to be blunted or, under some circumstances, completely absent. For instance, foreign tissues, rapidly rejected when placed into the skin, survive for a long time when placed into the CNS. This led to the concept of immune privilege of the CNS, i.e., the CNS was felt to be incapable of mounting any immune response, and whatever immune activity present was

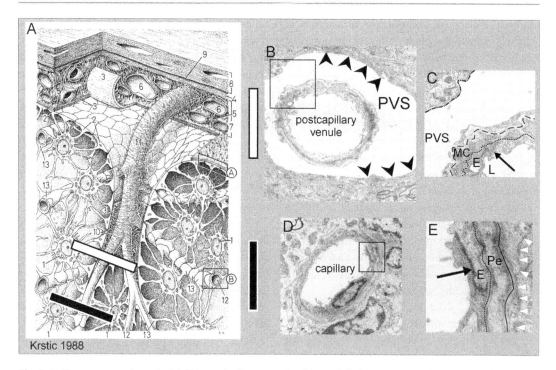

Fig. 3.1 The neurovascular unit. (**a**) Schematic diagram of different anatomic regions from brain parenchyma to skull. Superficial vessels of the brain [6] are located in the subarachnoid space [7]. This compartment is delineated by the arachnoid mater [4] and the pia mater [3]. The surface of the brain is completely covered by the astrocytic endfeet of the glia limitans [2]. Toward the subarachnoid space, these endfeet are designated as glia limitans superficialis (A); toward the vessels inside of the brain, they are termed glia limitans perivascularis (B). On their way from the surface to the deep areas of the brain, the vessels take leptomeningeal connective tissue with them, thereby forming perivascular spaces [10], which remain connected to the subarachnoid space. 1 indicates perikaryon of an astrocyte; 2 glia limitans superficialis; 3 connective tissue of the pia mater (inner layer of the leptomeninges); 4 arachnoid (outer layer of the leptomeninges); 5 subarachnoid connective tissue (trabeculae arachnoideae); 6 subarachnoid vessel; 7 subarachnoid space; 8 dura mater (pachymeninges); 9 neurothelium; 10 perivascular space; 11 penetrating vessel; 12 capillary; and 13 glia limitans perivascularis. (**b** and **c**) Ultrastructure of the PVS, which is bordered by other walls of a postcapillary venule and the glia limitans (*arrow-heads*) or their basement membranes. The field corresponds to the *white bar* (**a**). (**c**) Higher magnification of the field depicted (**b**) shows the basement membranes. In noncapillary vessels, at least three basement membranes can be distinguished: the endothelial membrane (*dotted line*), the outer vascular membrane (*dashed line*), and the membrane of the glia limitans (*dotted* and *dashed*). (**d**) Ultrastructure of the region corresponding to the *black bar* (**a**). (**d**) In capillaries (12 in **a**), the basement membranes are merged to form a "fused gliovascular membrane" that occludes the perivascular space. (**e**) Higher magnification of the field depicted (**d**). The capillary wall consists of endothelium [E], endothelial basement membrane (*dotted line*), and Pe. The gliovascular membrane is shown by a continuous *black line*. It is directly apposed to the glia limitans. Astrocyte processes of the glia limitans are indicated by *arrowheads*. The *arrows* (**c** and **e**) point to tight junctions between endothelial cells. The overlap of adjacent endothelial cells is a hallmark of the BBB and is evident in the capillary but not in the venule. *E* endothelial cell, *L* lumen, *MC* mural cell, an intermediate form between smooth muscle cell and pericyte, *Pe* pericytes, *PVS* perivascular space

completely dependent on cells recruited into the CNS. However, newer concepts are emerging, in which it is clear that the immune response in the CNS is simply different, and that the term "immune privilege" is misleading. For instance, the trafficking of dendritic cells, cells critical for adaptive immune responses (see Chap. 1) within the CNS, is very different within the CNS from that of other tissues, and their preferential homing to B-cell follicles in cervical lymph nodes may be responsible for the known "B-cell dominance" of immune responses within the CNS [4]. In addition, neurons and glial cells likely regulate the function of macrophages and lymphocytes once they enter the CNS [5]. Thus, the perceived "privilege" is one of active regulation, rather

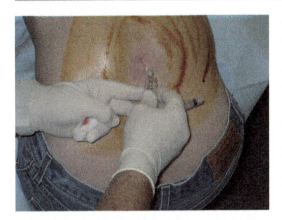

Fig. 3.2 Positioning for a lumbar puncture. The patient can be either in the lateral decubitus position, or in the sitting position, as in this patient. The needle is inserted into the subarachnoid space midline in the interspace between the fourth and fifth lumbar vertebrae, at the level of the iliac crest

than passive absence, not surprising given the dangers that unregulated immune responses would pose to the critical communications and control systems within the CNS [6].

2 Semple Rabies Vaccine Autoimmune Encephalomyelitis: Temporal Progression. Interplay Between the Nervous and Immune Systems

The development of antigen-specific T and B cells from lymphocyte precursors can only occur in lymphoid tissue. A good initial illustration of the interplay of peripheral lymphoid tissue and the CNS can be found in the "classical" neuroimmunological disease, Semple rabies vaccine autoimmune encephalomyelitis (SAE), as the disease proceeds in stages.

2.1 Stage 1. DAY 0: Exposure to the Antigen

The onset of many neuroimmunological and neuroinfectious diseases can be traced to a specific inciting event that disrupts homeostasis. In SAE, the inciting event is the injection of the human

with Semple rabies vaccine. For most antigens, the route is important since antigen processing is different at different sites of entry. In SAE, the route is subcutaneous, so lymph node drainage (see below) is to lymph nodes draining the skin in areas of injection. Entry of antigens or pathogens can be by a variety of other routes the most common being gastrointestinal and respiratory. These entry points have their own highly developed lymphoid tissue, called gut-associated lymphoid tissue (GALT) and bronchus-associated lymphoid tissue (BALT) which are specialized to deal with antigens coming in via these routes.

2.2 Stage 2. Days 0–7: Lymph Node Drainage and Processing of Antigen in Peripheral Lymph Node

Semple vaccine is injected subcutaneously at multiple sites, frequently in the abdomen, so the antigen would be initially present initially in the subcutaneous fat, and then be drained into draining lymph nodes including those in the inguinal area. As described above, Semple vaccine is not a pure substance, but a mixture of materials containing antigens including those in the rabies virus as well as CNS antigens such as myelin basic protein (MBP) [7] from the sheep brain material in the vaccine. These rabies and brain antigens are "processed" by dendritic cells and other "antigen-presenting" cells, mechanisms which begin locally, i.e., subcutaneously, and continue in the lymph node. The local lymph node is the site where the antigen-containing dendritic cells home usually via lymphatic vessels. During the first week, antigen-specific lymphocytes, B cells and T cells, including myelin-specific cells, are being activated, leave the lymph node via lymphatic vessels, enter the blood via the thoracic duct, and start circulating in the bloodstream. A particularly important subset of T cells in the early response is the Th17 cell, a helper T cell which secretes IL-17. The differentiation and expansion of this T-cell subset is favored by the presence of the cytokines transforming growth factor-β (TGF-β) and IL-6.

2.3   Stage 3. Days 7–10: Recruitment of Myelin-Specific Lymphocytes into the CNS

Lymphocytes normally traffic throughout the body and only stop either in secondary lymphoid organs or in tissues in which they encounter antigens with which their antigen receptors react. As part of the usual lymphocyte trafficking through the CNS, lymphocytes traffic through the CNS at low levels, i.e., under normal circumstances, lymphocytes enter, and then leave the CNS because there are no antigens with which the lymphocytes interact. However, in SAE due to the myelin present in the immunizing material, there are newly formed myelin-specific lymphocytes. These lymphocytes stop within the CNS, and interact with myelin antigens. In contrast, those newly formed T cells recognizing rabies antigens in Semple vaccine continue to circulate since there are no rabies antigens in the CNS in a vaccinated individual. T cells which circulate through the CNS are presumed to be effector memory cells, which express little or no CCR7 (a chemokine receptor associated with cell trafficking in the lymph node), and which secrete cytokines and express cytokine receptors. Less is known about B cells which traffic through the CNS, but the CNS is thought to be "B-cell friendly" because of the constitutive production within the CNS of survival factors for B cells [8]. The cells which traffic into the CNS are only a very small percentage of total lymphocyte trafficking. There are about 500 billion lymphocytes at any time in the human body, and about the same number traffic through the blood each day and go to and through other lymphoid tissues. At any time, most lymphocytes are found in lymph nodes, spleen, bone marrow, thymus, and mucosa-associated lymphoid tissue with only a very small number in tissues or in blood at any time [9]. Cells which contain class II MHC antigens, such as macrophages and dendritic cells from the blood, as well as microglia within the CNS, present MBP and other CNS antigens to antigen-specific T and B cells trafficking through, and these cells become locally activated, and go through cycles of replication. The release within the CNS

of cytokines by these myelin-specific cells activates other cells locally; an important consequence of this activation and cytokine release is that the blood–brain barrier is compromised in areas of inflammation.

2.4   Stage 4. Days 10–20: Maximal Inflammation with Involvement of Local CNS Immunity, Including Cervical Lymph Nodes

In this fourth stage of SAE, the processes outlined in stage 3 expand in amplitude and occur simultaneously, in multiple areas of the CNS, in the brain, brain stem, and spinal cord, due to trafficking and replication of myelin-specific lymphocytes. Symptoms of CNS dysfunction appear, such as weakness, numbness, tingling, and problems with vision, balance, bladder and bowel function, speech, or the cranial nerves. Cerebrospinal fluid (CSF) analysis performed during this stage reveals many white cells in the normally clear CSF. The CSF will also contain anti-Semple vaccine antibodies, produced by multiple plasma cells, each one making thousands of molecules of a clonal anti-Semple vaccine antibody. The precise mechanism of injury/dysfunction in SAE is not known, i.e., we do not really understand how immune activation and inflammation leads to CNS dysfunction. Many parts of the CNS can have considerable amounts of inflammation without resultant clinical sequelae.

As cells within the CNS die, and dendritic cells, macrophage, and other similar cells are recruited into the CNS from the blood, they encounter a range of breakdown products of the CNS from the injury. These will include neural antigens present in the original Semple vaccine, but in addition new antigens of human, not sheep or goat origin. Since there will be cross-reactions because of the similarity of molecules, these molecules will likely amplify the inflammation and extend its duration.

The precise manner in which tissue debris is cleared from the CNS, and inflammatory cells traffic, is unknown. Experiments in rats in the late 1990s had demonstrated that immune cells from the brain traffic to cervical lymph nodes [10], and then into the circulation, but recent experiments have revealed that another route, i.e., directly into cerebral blood vessels, is also possible. The inflammation will be evident to the neurologist as the symptoms experienced by the patient, the abnormal neurological examination consistent with CNS damage, increased number of white cells and protein in the CSF after lumbar puncture, and abnormal signals of inflammation and loss of CNS myelin evident on imaging of the brain.

2.5 Stage 5. Days 20 and Later: Recovery

After the peak of inflammatory process in the second to third week, poorly understood downregulatory mechanisms kick in. It is possible that a mix of "regulatory" cells, including T cells, B cells, microglia, and macrophages, contribute to this downregulation. Cellular debris is cleared and T- and B-cell numbers in the CNS decrease. Demyelination moderates and then stops, and remyelination, which had begin to occur already early in Stage 4, predominates. Clinically, the patient's symptoms begin to improve. Patients who survive the sometimes very stormy first few weeks of SAE frequently have a good outcome, and many patients return to normal within a few weeks to months following the onset of the disease.

As mentioned above, SAE is closely related to the other monophasic inflammatory syndromes EAE and ADEM. In EAE, the inciting immune stimulus is known because it is an induced experimental model, while in the human disease ADEM, the inciting antigen is usually not known. The controversies about the relationships between the inflammation and monophasic course of ADEM and its experimental model EAE and the human disease multiple sclerosis will be discussed in depth in the chapters on multiple sclerosis. Both ADEM and EAE are monophasic diseases with complete or near complete recovery and no further CNS involvement, while MS is a relapsing and progressive disabling inflammatory disease.

3 The Tools of the Neuroimmunologist

1. *History.* From a carefully obtained history, most neurologists can localize where within the nervous system the lesion lies, as well as determine the most likely few causes of the problem. At University Hospital in Newark, where I practice neurology, in months when I attend on service, I consistently can identify the nervous system location and etiology of an injury in 90% of cases by history alone. The pace and severity of the symptoms and exacerbating/remitting factors are helpful. By probing for problems referable to certain locations, e.g., speech content problems coming from the left temporal lobe pathology, lesions can usually be accurately localized. A careful past medical history, social history, and family history frequently demonstrate that the patient being evaluated is at a high risk for certain disease processes.

2. *Examination.* A thorough neurological examination, as briefly outlined in Chap. 2, is necessary, and often will confirm the hypotheses drawn from the history. Most patients with multiple sclerosis, for instance, will have involvement of multiple areas of the nervous system, with problems with vision, eye movement, balance, tone, strength, and sensation, which can all be documented on the neurological examination.

3. *Blood analysis.* A complete blood count may show an increased number of white blood cells consistent with a generalized infection. An elevated erythrocyte sedimentation rate (ESR) would be consistent with a systemic vasculitis, such as giant cell arteritis. PCR for bacteria and viruses can be done in the blood, e.g., patients with AIDS have levels of the human immunodeficiency virus (HIV) in their blood.

4. *CSF analysis.* Most patients being considered for a neuroimmunological disease will undergo a lumbar puncture (LP) (Fig. 3.2) to obtain and test CSF. Similar to the immune system which has its own fluid (lymph) and its own circulation (through lymph vessels), the CNS

has its own fluid (CSF) and its own circulation, within the subarachnoid space around the brain and spinal cord. Inflammation within the CNS can usually be readily detected by CSF analysis. Some common processes producing CSF abnormalities were summarized in Table 2.2 in Chap. 2.

5. *Imaging of the CNS or PNS.* Computerized tomography (CT) or magnetic resonance image (MRI) is used to image the brain, spinal cord, or other areas of the body. Multiple sclerosis causes a characteristic picture on MRI scanning of the brain of loss of myelin in multiple areas of the CNS associated with breach of the blood–brain barrier, while, in herpes simplex virus encephalitis, there is one focal area of increased signal in the temporal lobe unilaterally. Inflamed areas of the brain can sometimes additionally be detected by injecting a dye named gadolinium intravenously during MRI scanning; because of the local breakdown in the blood–brain barrier in vessels of inflamed areas, gadolinium will appear in the parenchyma of the inflamed areas. This can aid in diagnosis and is commonly used especially in multiple sclerosis.

4 Aspects of Inflammation in Neuroimmunology Unique to the Nervous System

1. Variability in clinical outcome depending on the *location* within the nervous system of the inflammatory process. In inflammation of the lung or liver or other tissue, location within the tissue will generally not markedly affect the symptoms and signs of the illness. In neuroimmunological disease and in neurological disease in general, location within the CNS is extremely important, e.g., inflammation in the temporal lobe cortex such as that seen in herpes encephalitis, with seizures and headache, has a much different presentation than that seen in inflammation in the spinal cord, such as seen in myelitis from multiple sclerosis, when the patient loses strength in her legs and loses bladder function.

2. *Limited* access of inflammatory cells to the parenchyma: In most tissues, there is free access of cells from the circulation to an inflammatory process within the parenchyma; this is not true for the nervous system, where the blood–brain barrier and the blood–nerve barrier regulate influx of cells. The nervous system regulates these barriers by a variety of mechanisms. It adjusts the levels of adhesion molecules required for white cells to attach and enter. It modulates the tightness of the tight junctions of the endothelial cells. Astrocytes are tightly linked to the perivascular space and are thought to affect the movement of cells into the parenchyma.

3. *Absence* of anatomically defined lymphatic system: Most tissues are aided in their interaction with the immune system by the lymphatic system which consists of lymph vessels and lymph nodes. There is no anatomically defined lymphatic system in the CNS, although movement of CSF and interstitial fluid through the brain and Virchow-Robins spaces, i.e., the perivascular spaces, with subsequent drainage to cervical lymph nodes, subserves some of the role of lymphatics [11].

4. The *anti-inflammatory* "*tone*" of the CNS: A hallmark of inflammation is swelling, also called "tumor." Medical students for 20 centuries have learned the four cardinal signs of inflammation in Latin, first recorded by Celsus in the first century AD: calor (heat), dolor (pain), rubor (redness), and tumor (swelling). Swelling in immune-mediated processes such as delayed-type hypersensitivity (DTH) is T cell mediated. The brain in adults is encased by the rigid skull, so swelling in the brain leads to increased pressure and potentially fatal complications. This fact may account evolutionarily for the "anti-inflammatory" tone of the CNS. The precise molecular causes for what has been called "a hostile environment for T cells" in the CNS is not known, but high levels of cytokines such as TGF-β or IL-10 which are usually downregulatory for T-cell functions may be contributing. In contrast, swelling in nonneural tissues is generally not so damaging, and for these tissues, the

full range of inflammation, including tumor, is allowed. Since it is primarily Th1 and Th17 cells which mediate DTH, and thus swelling, it is not surprising that there is a relative skew in the CNS away from these T cells toward the function of Th2 cells [12], which help B cells make antibody.

5. *Microglia: A unique CNS immune cell.* Microglia, which are pluripotent cells unique to the CNS, make up approximately 10% of the glia in the CNS; their functions are poorly understood. Immune stimuli result in at least two changes in this CNS-resident cell population: activation and proliferation. They are cousins to monocytes and tissue macrophages, and they cannot be readily distinguished in inflamed CNS from macrophages infiltrating from the blood. In addition to being able to opsonize and process debris, they can also produce a number of cytokines, as well as serve as antigen-presenting cells. There is a evidence that they respond to signals from astrocytes and neurons.

6. *The HIV/AIDS revolution.* The current practice of neuroimmunology in the USA involves two distinct populations of patients for which different approaches must be taken: those who are infected with HIV and those who are not. The spectrum of diseases in each of these two groups is very different, so determination of HIV status is important particularly in patients with neuroimmunological disease which is difficult to diagnose.

At the end of 2006, over a million individuals in the USA were estimated to have HIV infection; in more than a fifth, the infection had not yet been diagnosed. HIV infection leads to deficits in the immune response, markedly increasing the probability of severe infections. Some of these infections, called opportunistic infections (OIs), are found normally in a controlled, low-level state in immunocompetent individuals, but, because of immunodeficiency in the HIV-infected patient, these pathogens proliferate, become uncontrolled, and cause injury or death. OIs are AIDS-defining illnesses; i.e., prior to getting these illnesses the patient is considered as simply having HIV infection. The most prominent neurological infections

among the OIs are cerebral toxoplasmosis, progressive multifocal leukoencephalopathy, and cytomegalovirus polyradiculitis. Tuberculous meningitis and varicella zoster virus complications, including shingles and myelitis, can also be seen. Fungal infections, predominantly with cryptococcus and also candida, aspergilla, and histoplasma, are also seen. Direct injury to the nervous system from HIV occurs with HIV dementia (also called AIDS–dementia complex or ADC), AIDS myelopathy, and HIV-associated neuropathy. These infections will be discussed in Chap. 12.

5 The Necessity for Great Care in Classifying a Neuroimmunological Disease as "Autoimmune"

Many neuroimmunological diseases have been classified by some as "autoimmune" because they are associated with lymphoid cell infiltrates and because a pathogen is not an obvious cause of the problem. This, of course, is inaccurate, and potentially dangerous. Most neuroimmunological diseases are idiopathic, i.e., their etiology and pathogenesis are unknown. Calling a disease "autoimmune" indicates that the disease process is an attack by the immune system on an identified nervous system antigen or combination of antigens. This concept has recently been summarized by Moses Rodriguez in an editorial in *Annals of Neurology* manuscript titled "Have We Finally Identified An Autoimmune Demyelinating Disease?" [13]; he also summarized criteria for an autoimmune disease (Table 2.2), and identified how NMO meets most of these criteria. Even if an autoantibody can be identified in a disease, it is not necessarily indicative that the disease is truly autoimmune, since autoantibodies can be induced by infections, e.g., a common autoantibody used in testing for syphilis is the anti-cardiolipin autoantibody, i.e., the RPR or VDRL tests (see discussions of testing for syphilis in Chaps. 12 and 18). No one would consider syphilis an autoimmune disease, but that is because the pathogen of syphilis has been identified. It is likely that many of the diseases which some are

Table 3.2 Criteria for autoimmunity (modified from Rodriguez [13])

Criteria	Neuromyelitis optica	Myasthenia gravis	Multiple sclerosis
Demonstration of an immune response to a precise autoantigen in all patients with the disease	Yes. AQP4 (but not all patients)	Yes. AChR or MUSK in large percentage	No
Reproduction of the lesion by administration of autoantibody or T cells into a normal animal	Yes	Yes	No
Induction of lesions by immunizing an animal with relevant purified autoantigen	Not done	Yes	No
Isolation or presence of autoantibody or autoreactive T cells from lesion or serum	Yes. Anti-AQP4 antibodies in the serum and pathological lesions	Yes	No
Correlation of autoantibody or autoreactive T cell with disease activity	Yes	Yes. Prospectively, within the same patient	No
Presence of other autoimmune disorders or autoantigens associated with disease	Yes	Yes	No. Slight increase in autoimmune disease prevalence compared to control population
Immune absorption with purified autoantigen abrogates pathogenic autoantibody or autoreactive T cell	Not done	Yes	No
Reduction of pathogenic autoantibody or T cell associated with clinical improvement	Yes. Plasma exchange	Yes	No

incorrectly identifying in the present as "autoimmune" will in the future turn out to be infectious in pathogenesis.

The classification of a disease as being autoimmune has important diagnostic and therapeutic ramifications. Thus, in diseases with likely autoimmune pathogenesis such as NMO and myasthenia gravis (MG), (Table 3.2), the diagnosis will be supported by demonstration of the autoimmune process, such as antibodies to aquaporin 4 or the nicotinic acetylcholine receptor (AChR). In contrast, in diseases which are likely not autoimmune, an immune response to a neurological antigen cannot be used to aid in diagnosis. In autoimmune diseases such as NMO and MG, immunosuppressive medications, some of which have substantial associated risks, might be used aggressively, while in diseases likely not to be autoimmune, the extended use of potent immunosuppressives would be less likely to benefit the patient. Similar care must be used when the words "inflammatory" or "immunologically mediated" are used, since these adjectives are often used indiscriminately.

6   The Importance of Antibodies

Antibodies, the product of B lymphocytes, are the only molecules of the immune response that can interact with high affinity with a foreign molecule in solution. They are the critical central molecule in the humoral immune response. They have become highly important molecules to the neuroimmunologist in many ways outlined below. In contrast, the T lymphocyte, the central cell of the adaptive cellular immune response, has not been able to be utilized to a similar extent in the practice of clinical neuroimmunology. Despite its clear importance in basic research on neuroimmunological disease, measurement of T-cell function or use of T-cell-derived molecules has not significantly entered clinical practice with the exception of measurement of T-cell subsets in patients suspected of being infected with the HIV.

First, a brief clarification regarding nomenclature: the words "immunoglobulins" and "antibodies" are generally used interchangeably. Frequently "immunoglobulin" refers to molecules which have structures of immunoglobulin G, M, or the other isotypes, but whose specificity for any specific antigen is unknown. Conversely, "antibody" thus usually is the word used for an immunoglobulin molecule whose antigenic specificity is known. As an example, in the anti-Semple vaccine immune response described earlier in the chapter, the anti-Semple vaccine antibodies produced in this response are immunoglobulin molecules with high affinities for molecules present in the Semple vaccine.

The importance of antibodies in neuroimmunology stems from their extensive use both diagnostically and therapeutically. These are summarized briefly below:

1. Antibodies used for diagnostic purposes:
   (a) *Infections*: For many infections, infection-induced antibodies assist in diagnosis of the infection. In fact, for infections in which the causative organism is difficult to grow in culture or to amplify by PCR, anti-pathogen antibodies in the blood or CSF are almost the only laboratory test which can assist the clinician. Examples discussed in Chap. 12 are Lyme disease, syphilis, and West Nile virus.
   (b) *Autoimmune diseases in which the molecular target is known*: Myasthenia gravis or NMO, described in Chaps. 6 and 9, is diseases in which the autoimmune process generates antibodies to the anti-AChR or antibodies to aquaporin, which aid in diagnosis.
   (c) *Paraneoplastic autoantibodies*: In these diseases, described in Chap. 14, which cannot be characterized as autoimmune using standard criteria, antibodies reactive to identified CNS proteins can be used as biomarkers of the disease. An example is anti-Purkinje cell antibodies in paraneoplastic cerebellar degeneration.
   (d) *Therapy induced*: Some forms of therapy for neuroimmunological disease induce antibodies to the therapy. Examples are anti-interferon-β or anti-natalizumab antibodies in multiple sclerosis therapy. These therapies, and their associated anti-therapy antibodies, are described in Chap. 7. They often are called neutralizing antibodies (NAbs), since they neutralize the therapeutic effect of the drug.
2. *Therapeutic*:
   Since 1986 when the first one was approved by the FDA, therapeutic monoclonal antibodies

have become an exponentially growing market. Twenty-two monoclonals are now approved by the FDA for therapy and there are hundreds more in clinical trials. One of the top selling and most established monoclonal antibodies, approved by the FDA in 1997, is rituximab, targeting CD20, which is in clinical trials for therapy of MS. Natalizumab, rituximab, alemtuzumab, and daclizumab are described in greater depth in Chaps. 7, 17, and 19.

References

1. Hemachudha T, Phanuphak P, Johnson RT, Griffin DE, Ratanavongsiri J, Siriprasomsup W. Neurologic complications of Semple-type rabies vaccine: clinical and immunologic studies. Neurology. 1987;37(4):550–6.
2. Waksman BH. A history of neuroimmunology; a personal perspective. In: Antel J, Birnbaum G, Hartung HP, Vincent A, editors. Clinical neuroimmunology. 2nd ed. Oxford: Oxford University Press; 2005. p. 425–45.
3. Hohlfeld R. Multiple sclerosis: human model for EAE? Eur J Immunol. 2009;39(8):2036–9.
4. Hatterer E, Davoust N, Didier-Bazes M, et al. How to drain without lymphatics? Dendritic cells migrate from the cerebrospinal fluid to the B-cell follicles of cervical lymph nodes. Blood. 2006;107(2):806–12.
5. Carson MJ, Doose JM, Melchior B, Schmid CD, Ploix CC. CNS immune privilege: hiding in plain sight. Immunol Rev. 2006;213:48–65.
6. Pachner AR. The immune response to infectious diseases of the central nervous system: a tenuous balance. Springer Semin Immunopathol. 1996;18(1):25–34.
7. Piyasirisilp S, Hemachudha T, Griffin DE. B-cell responses to myelin basic protein and its epitopes in autoimmune encephalomyelitis induced by Semple rabies vaccine. J Neuroimmunol. 1999;98(2):96–104.
8. Meinl E, Krumbholz M, Hohlfeld R. B lineage cells in the inflammatory central nervous system environment: migration, maintenance, local antibody production, and therapeutic modulation. Ann Neurol. 2006;59(6): 880–92.
9. Westermann J, Pabst R. Distribution of lymphocyte subsets and natural killer cells in the human body. Clin Investig. 1992;70(7):539–44.
10. Knopf PM, Harling-Berg CJ, Cserr HF, et al. Antigen-dependent intrathecal antibody synthesis in the normal rat brain: tissue entry and local retention of antigen-specific B cells. J Immunol. 1998;161(2):692–701.
11. Hickey WF. Basic principles of immunological surveillance of the normal central nervous system. Glia. 2001;36(2):118–24.
12. Harling-Berg CJ, Park TJ, Knopf PM. Role of the cervical lymphatics in the Th2-type hierarchy of CNS immune regulation. J Neuroimmunol. 1999;101(2):111–27.
13. Rodriguez M. Have we finally identified an autoimmune demyelinating disease? Ann Neurol. 2009;66(5):572–3.

The Prototypic Neuroimmunological CNS Disease: Multiple Sclerosis, a Precis

4

"Can you do Addition?", the White Queen asked. "What's one and one and one and one and one and one and one and one and one and one?"
"I don't know," said Alice. "I lost count."

Lewis Carroll, *Through the Looking Glass*

1 Definition

Multiple sclerosis (MS) is a disease of the central nervous system primarily affecting young adults, usually characterized by the development of sudden exacerbations (attacks) with remissions of focal neurological dysfunction and the development of progressive neurological disability over time. Lesions in the CNS are continually accrued ("one and one and one…") so that over time, the CNS is damaged as a function of the frequency and location of lesions balanced by as yet poorly understood reparative processes. It is characterized pathologically by inflammation and demyelination as well as remyelination exclusively in the central nervous system, with no pathology in the peripheral nervous system or non-nervous system tissue. Its cause is unknown and its treatment is currently suboptimal in that exacerbations and disease progression are not universally halted. It is a uniquely human condition and although there are animal models, none is fully faithful to the human disease. In some patients with MS, those with the primary progressive form of the disease, there is simply progression of disability without exacerbations or remissions.

2 Etiopathogenesis

The etiology and pathogenesis of MS are unknown. Many investigators through the decades have been very convinced that their particular hypothesis is correct, but the etiology remains enigmatic. There is some genetic proclivity since family members of MS patients have higher incidences of MS relative to the general population. These aspects will be described in greater detail in the section on genetics and epidemiology of MS (below).

The history of "changing paradigm shifts" in MS pathogenesis has been well-summarized by Byron Waksman [1]. In the nineteenth century, during the first years of neurology as a field of medicine, "disseminated sclerosis" was thought to be due to glial scarring. Later, MS was thought possibly to be a thromboembolic phenomenon. In the early years of the twentieth century, MS was considered as a problem of myelin breakdown, and later in the twentieth century, measles virus was an attractive causative agent. It was not until immunology emerged as a major field of scientific endeavor in the latter third of the twentieth century that neurologists became interested

A.R. Pachner, *A Primer of Neuroimmunological Disease*,
DOI 10.1007/978-1-4614-2188-7_4, © Springer Science+Business Media, LLC 2012

in autoimmunity as a possible cause. Waksman stressed that the increased interest in a possible immune pathogenesis went hand-in-glove with advances in immunology, arguably the most rapidly moving field in medicine over the last 50 years. One of the reasons that paradigm shifts have occurred and likely will continue to occur is that, although each postulated pathogenesis has sounded exciting and interesting, none of them has yet fully fit the experimental data or been able to predict responses to therapies. The fullness of time will tell whether the autoimmunity hypothesis has a more lasting tenure than did glial scarring, thromboemboli, myelin breakdown, or measles virus. Currently, the most prominent hypotheses remain autoimmunity and chronic infection, theories which are not mutually exclusive.

Another important summary of the many hypotheses that have come and gone can be found in Murray's excellent book, *Multiple Sclerosis: The History of a Disease*, especially Chapter 11, "Searching for a Cause of MS." These very well-written 90 pages provide the reader with a good understanding of the search for an infectious, genetic, epidemiological, immunological, vascular, or environmental cause.

Currently, there are camps representing some of the leading hypotheses for pathogenesis of MS. The largest camp consists of those scientists and clinicians who are convinced that the problem is "autoreactive T cells," thus an autoimmune disease. This group is opposed by an equally impressive group who feels that the "autoreactive T cell" hypothesis remains just that: a hypothesis which is largely unproven despite the expenditure of a great amount of funding, time, and effort in that direction, and that the cause of the disease lies elsewhere, such as an infectious pathogen or a neurodegenerative process.

There are cogent arguments on both sides, very briefly reviewed below. Supporting the "T-cell autoimmune" hypothesis are animal experiments in which, autoreactive T cells can cause inflammation in the brain, and the pathology shows T cells within lesions. Therapies which affect the immune response, such as corticosteroids, IFN-β, and natalizumab (a humanized anti-integrin antibody) have an ameliorative effect on certain aspects of the disease. On the possible pathogen side are many arguments against T-cell mediation such as the absence of the effect of the anti-T-cell drug cyclosporine [2], the failure to demonstrate significant autoimmune T-cell activity in MS patients, and the fact that several viruses reproduce the features of MS when injected into experimental animals (see Sect. 1 of Chap. 8).

These arguments have more than just academic repercussions. For many "T-cell autoimmunity" proponents, therapy of MS requires aggressive immunosuppression using medications or interventions that carry significant risks and have unproven efficacy over an extended period of time. Those who do not believe in the "T-cell autoimmunity" hypothesis favor using safer medications and supporting research into nonimmune aspects of the disease to identify nonimmune targets for therapy. The situation is made more complex by the fact that the number of relapses does not correlate well with progressive disability (see Chap. 6), indicating that the primary measure which the FDA uses to judge therapies, i.e., relapse rate, may be irrelevant for the aspect of the disease that is most damaging, i.e., disability progression. Unfortunately, given these current complexities of the disease and its treatment, the lack of well-run long-term studies of the various agents, and controversy about the etiopathogenesis of the disease, the use of most of these medications over the long term to substantially improve the natural history of the disease relative to disability remains more "faith-based" rather than "evidence-based."

3   Pathology

The pathology of multiple sclerosis is both distinctive and unique. The most obvious pathology, easily seen in gross as well as microscopic specimens, is the MS plaque, representing well-circumscribed loss of myelin (see Chap. 2). Late nineteenth century neurologists, especially Charcot and others at the Saltpetrière, included autopsy in their repertoire of learning about the nervous system. They commented on the feel of the areas in which myelin was lost, the firmness of the tissue in contrast to the soft normally myelinated tissue around the plaque. This prompted the name in French, "sclérose en plaques disseminée," and in France today, the term "sclérose en plaques" (SP) is still used. "Multiple sclerosis" became the term which achieved fairly broad international acceptance for the condition in the 1950s, especially after the publication in 1955 of McAlpine's classic text on the disease bearing that name as its title [3].

In early MS, there is prominent perivenular inflammation and little, if any, atrophy. As the disease progresses, there are more and more "sharp-edged plaques" [4], areas of near complete loss of myelin with a sharp border surrounded by normally myelinated tissue. Areas of atrophy, initially small, become confluent and there are increased numbers of demyelinated plaques, which also become confluent. Inflammatory cells, including plasma cells, become more prominent within the parenchyma. Sharp-edged plaques had for years been thought to be unique for MS, but have relatively recently been described in a variation of the EAE model in the marmoset [5], using a protocol of immunization with human white matter (see Chap. 8).

The pathology of MS prior to 1996 was felt to be relatively homogeneous. Then, in 1996 Lucchinetti et al. postulated four different forms of MS lesions, with only one of the forms occurring in any one patient, which they felt had pathogenetic significance [6] and signified "different immunological mechanisms of myelin destruction in MS." The subsequent years have seen considerable resistance to this classification, and a recent paper identifies "a uniform pre-phagocytic pathology and overlap of lesion subtypes in individual patients with typical relapsing and remitting disease" [7]. It remains controversial whether there are substantially different pathologies or even different biologies of MS.

4 Genetics and Epidemiology

The prevalence of MS is highly variable depending on geographic region. It has long been thought that countries near the equator have a low prevalence, with increasing prevalence moving either north or south from the equator. There is no accepted explanation for this. However, some recent studies from Latin America do not support this gradient of MS prevalence.

Individuals with MS in the family are more likely to develop the disease than individuals who lack this family history, and the likelihood increases the closer the genetic relationship to the proband. Ebers [8] has summarized the data recently in an excellent review. In a population in which the prevalence of MS is 1/1,000, relatives have prevalences as follows: adoptive siblings—1/1,000, first cousin—7/1,000, full sibling—35/1,000, and monozygotic twin—270/1,000. That no single gene is highly correlated with the disease indicates the polygenic basis for MS. The nearest is the link to the major histocompatibility complex (MHC), especially to class II MHC. However, associations with the MHC are complex, and the associations are not as straightforward as previously thought [9].

5 Clinical Manifestations

5.1 Initial Symptoms

Many of the first symptoms of MS are so minimal and fleeting that they are ignored. There may be some numbness or tingling in an arm or leg for a few days with resolution, followed by months or years of normal neurological health. Then a few weeks of problems with balance attributed to a "virus" may occur, again with resolution, followed by no symptoms for years. Finally, blindness in one eye prompts an investigation leading to the correct diagnosis. In the majority of patients with MS, at their first diagnosis there is clinical and MRI evidence, suggesting that the disease had been present years prior to their diagnosis.

Multiple sclerosis means "many plaques," and the lesions, and thus clinical symptoms, can be anywhere there is CNS myelin. CNS myelin is present in all white matter in the CNS, and areas which are particularly frequently targeted are periventricular areas, optic nerves, cerebellar and pontine white matter, and white matter tracts in the cervical and thoracic spinal cords. Little is

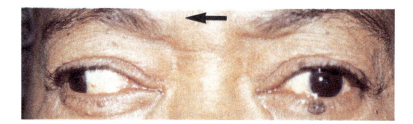

Fig. 4.1 Internuclear ophthalmoplegia (INO): This patient is being asked to look to his right. His right eye is able to do that but not his left. A similar problem occurs when asked to look to his left

known of the factors which determine whether a given white matter lesion will result in any clinical sequelae. In a recent MS study using monthly MRI scans with high sensitivity for new MS lesions [10], for every new MRI lesion leading to symptoms, there were 17 new MS lesions that did not result in any clinical symptoms, a number consistent with other studies in the literature.

5.2   The MS Attack

> …this girl's clinical picture can NOT be produced by one lesion. It is like the mathematics lesson in *Through the Looking Glass*, a process of addition. No one sign constitutes the diagnosis; it is the sum of many over the years. There is manifest evidence of damage to the cerebrospinal motor pathways. If a preparation of the brain and spinal cord were nailed to a barn door and shot at with a spread of No.6 shot, the pyramidal tracts could not in all verity escape. And seldom are there lacking signs of such involvement in this disease. It consists of multiple, disseminated plaques of hardening in the spinal cord.
>
> Harry Lee Parker, M.D., "Multiple sclerosis" in *Clinical Studies in Neurology*

The archetypal presentation of MS is an "attack" or "relapse." There is the rapid development over hours to a few days of a neurological deficit referable to injury to some area within the CNS. An arm may go numb, double vision develop, one side of the face droop, or a leg become weak. The problem lasts for a few days to a few weeks and then slowly gets better and frequently completely resolves, even without therapy. Some syndromes are so characteristic that when they first appear in a young patient neurologists become particularly concerned about the possibility of MS. One of these is involvement of the medial longitudinal fasciculus (MLF), a white matter tract in the midbrain which provides for proper communication between the nuclei controlling eye movements so that conjugate eye movement can be performed. Lesions in the MLF cause a particular form of "eye weakness" (ophthalmoplegia), called "internuclear." Internuclear ophthalmoplegia (INO) results in symptoms of double vision, especially when looking to the side (Fig. 4.1). Another common initial manifestation is acute inflammation of the optic nerve (optic neuritis), which results in eye pain and loss of visual acuity, sometimes so severe as to cause complete blindness in that eye.

There is no universally accepted definition of an attack. For clinical trials in which the primary endpoint is number of attacks, a rigorous definition of an attack is needed, but they are defined differently in different trials. Most clinical trials will require a neurologist-documented worsening in the patient's neurological function, often linked to the EDSS (Expanded Disability Status Scale), sometimes called the Kurtzke score, a standard disability measure (see below). As the definition of relapse has become more important in the development of new drugs for MS, definitions have become more complex. An example of a definition of an "attack" or "relapse" can be found in a recent head-to-head study of two MS treatments, the BECOME study, which utilized two different MS rating scales and distinguished between a subjective and an objective relapse [10]: "All new or worsening neurologic symptoms lasting greater than 24 h and not explained by fever or infection were considered subjective relapses. Subjective relapses that were confirmed by a blinded examining neurologist using worsening scores on either the Scripps Neurological Rating Scale or the EDSS were considered objective relapses. One or more of the following changes compared with baseline was required for relapse confirmation: (1) increase in total EDSS by greater than or equal to 0.5 point; (2) increase in the EDSS score for one system greater than or equal to 2 points; (3) increase in the score of 2 or more EDSS systems greater or equal to 1 point; and (4) decrease in SNRS score by greater or equal to 7 points." However, in routine neurological practice neurologists usually will use their clinical experience to identify an attack. When specifically asked to define an attack, they might say, "I know one when I see it," and they are usually proven correct by the subsequent course of the patient.

After a first attack the patient is considered to have a "clinically isolated syndrome" (CIS) which may not end up being MS. Thus, a substantial percentage of patients with optic neuritis do not develop MS. Depending on the study, that number is 20–50%. The probability of developing "clinically definite MS" (CDMS) over time is substantially increased in CIS patients if they have MRI abnormalities consistent with MS outside of the areas in the CNS associated with their symptoms. Thus, a young otherwise healthy patient with optic neuritis is much more likely to develop MS if she is found to have periventricular white matter lesions on brain MRI than if no lesions are evident. This will be discussed at greater length in the next chapter.

As the disease continues and there is more involvement of the white matter, there is also axonal injury. Axonal injury has received much less attention than has demyelination for two reasons. First, the usual tool clinicians use to measure damage to CNS structure is magnetic resonance imaging. Because water constitutes a low percentage in CNS myelin and increases when that myelin is lost, the gain in the water signal is a very strong marker of demyelination. There is no equivalently straightforward way to

measure axonal injury on MRI. In fact, there is no accepted measure of axonal injury on MRI, although a number of excellent investigators are working on this problem. Second, the pathologic hallmark of MS has always been demyelination because myelin is easy to stain, and the dramatic loss of myelin is so obvious on pathology. A similarly obvious measure does not exist for axonal injury. When axonal injury has been studied, it appears to be a critical component of CNS injury in MS [11].

It is the combination of accrual over time of both axonal injury and demyelination with inadequate recovery that ultimately leads to the development of disability. Disability in MS can be measured in a number of ways, but the most commonly used measure in clinical trials is the EDSS [12] (Table 4.1). This scale involves the measurement of the amount of functional limitations in a number of neurological functions, which are factored in and collated to provide a single number. A score of zero indicates no disability and a score of ten is death from MS; a score of six is given to an MS patient who requires an aid to ambulate.

When patients see their physicians and when physicians weigh the efficacy of therapy, it is the rapidity of disability progression that is uppermost in their minds. The rapidity of accrual of disability over time is highly variable and is a feature of MS which can be called MS severity. MS severity can be measured with the MS Severity Score (MSSS) [13]. For instance, a patient with relatively mild disability, an EDSS of 3, would have a very low MSSS, if she has had the disease for 30 years. In contrast, a different patient with the same EDSS developed over 2 years will have a high severity score, since a substantial amount of disability was accrued over a very short period of time.

This accrual of neurological disability over time does not correlate with many other features of the disease. For instance, the number or severity of clinical attacks or the degree of CNS demyelination by MRI does not correlate very well with neurological disability. Since we have learned from MRI studies that there are many new MRI lesions for each new clinical attack, it is

Table 4.1 EDSS (also called Kurtzke score)

0	Normal neurologic exam
1.0	No disability, minimal signs in one functional system
1.5	No disability, minimal signs in more than one functional system
2.0	Minimal disability in one functional system
2.5	Minimal disability in two functional systems
3.0	Moderate disability in one FS, or mild disability in three or four functional systems though fully ambulatory
3.5	Fully ambulatory but with moderate disability in three or four functional systems
4.0	Fully ambulatory without aid, self-sufficient, up and about some 12 h a day despite relatively severe disability. Able to walk without aid or rest some 500 m
4.5	Fully ambulatory without aid, up and about much of the day, able to work a full day, ma otherwise have some limitation of full activity or require minimal assistance, characterized by relatively severe disability. Able to walk without air or rest for some 300 m
5.0	Ambulatory without aid or rest for about 200 m; disability severe enough to preclude full daily activities (e.g., to work full day without special provisions)
5.5	Ambulatory without air or rest for about 100 m; disability severe enough to preclude full daily activities
6.0	Intermittent or unilateral constant assistance (cane, crutch, or brace) required to walk about 100 m with or without resting
6.5	Constant bilateral assistance (canes, crutches, or braces) required to walk about 20 m without resting
7.0	Unable to walk beyond about 5 m even with aid. Essentially restricted to a wheelchair. Wheels self in standard wheelchair and transfers alone. Active in wheelchair about 12 h a day
7.5	Unable to take more than a few steps. Restricted to wheelchair. May need air to transfer. Wheels self but cannot carry on in standard wheelchair a full day, May require a motorized wheelchair
8.0	Unable to walk at all, essentially restricted to bed, chair, or wheelchair but may be out of bed much of the day. Retains many self-care functions. Generally has effective use of the arms
8.5	Essentially restricted to bed much of the day. Has some effective use of arm(s). Retains some self-care functions
9.0	Helpless bed patient. Can communicate and eat
9.5	Totally helpless bed patient. Unable to communicate effectively or eat/swallow
10	Death due to multiple sclerosis

Source: Kurtzke [12]

not surprising that the clinical attacks would not predict clinical disability very well, since the vast majority of MS-induced injuries to the CNS are subclinical. This has led to recommendations for the more frequent use of MRI outcomes for studies rather than using attack rate [14]. However, MRIs do best when CNS demyelination is being measured and even the extent of CNS demyelination does not correlate well with clinical disability. Clinical disability may be determined to a large extent by axonal injury, which is very difficult to measure by MRI. There is some evidence that disability correlates well with the development of spinal cord atrophy [15]. In the USA, pharmaceutical companies have used relapse data rather than disability data to obtain

FDA approval, of new treatments for MS, which is a problem since relapse frequency does not correlate with disability progression (see Chap. 7).

6   Natural History and Prognosis

The natural history of MS has been well documented for populations. For most patients, those with relapsing–remitting MS (RRMS), there are at least five stages (Table 4.2): the first is a "preclinical stage" in which white matter lesions detectable on MRI accumulate, but there are no symptoms identified by the patient. The second stage begins with the first development of clinical symptoms, usually an attack, which prompts the

Table 4.2 The five temporal stages of MS

Preclinical
First attack—clinically isolated syndrome (CIS)
Relapsing-remitting MS
Secondary progressive MS
Late MS

diagnosis of CIS. At some point, months or years later, further clinical or MRI lesions develop, resulting in the diagnosis of MS, and the beginning of the third stage, that of RRMS. In most MS patients, these attacks occur periodically, but early on, disability is not a dominant factor. At some point, relapses become less common and disability is more prominent and progressive. This stage is called secondary progressive MS (SPMS). In many patients, there is subsequently a fifth stage, in which disability plateaus and progression ceases. This temporal classification applies to patients diagnosed with RRMS.

A substantial portion of MS patients never have a true attack and simply have progressive disability over time, a condition called primary progressive MS (PPMS). Many patients have something in between, i.e., just one or two clinical attacks over time, followed by progressive disability; these patients will usually be classified as having RRMS.

As well as this natural history applies to populations, it applies poorly to the individual MS patient. When I see an early MS patient, I stress that each individual has his or her own form of MS: that Ms. X has the Ms. X form of MS, a very different form than Mr. Y or Mrs. Z. Suffice it to say that the vast majority of MS patients first see a neurologist when they are young, and 20 years later, by the time they are middle-aged, many have accrued a substantial amount of neurological disability. Not only is the disease variable from patient to patient, but it is also variable within any particular patient. Ms. X can have a flurry of activity in her early 30s, i.e., multiple attacks and the development of disability, and then again in her late 40s, with 15 intervening years of no clinical or MRI activity. The highly unpredictable nature of the disease creates great difficulty for any clinician who attempts to

provide the MS patient with a prognosis, and very few neurologists will attempt to predict the course of any particular patient.

7   MS Clinical Classifications

Given the complexities of MS, with its unknown etiology and unpredictable course, there have been many attempts at classification to provide some framework for understanding the disease (Inset 4.1).

Inset 4.1 Two Clinical Vignettes of Patients with MS

Patient #1. Mother of three with confusion and trouble walking
ST was a 27-year-old mother of three when she was seen at University Hospital for a second opinion.

She was completely well until 5 months prior to her visit. At that time, she developed visual problems and gait unsteadiness. An evaluation at that time was consistent with MS. This seemed to improve but very shortly thereafter she worsened and began to have personality changes, increasing problems with gait, weight loss, worsening speech problems. She was begun on interferon-β but continued to have worsening.

MRI scans of the brain over the past months prior to the first visit revealed multiple enhancing lesions distributed throughout her brain.

Examination of her first visit to University Hospital revealed a thin young woman in no acute distress. On mental status examination she was disoriented to time (November 1994—didn't know day), but oriented to place and person. She was barely able to repeat 3/3 objects, and her recall after 1 min was 0/3. Serial 3 s was performed down to 97, but no further. She knew the names of the President the Vice President.

(continued)

Inset 4.1 (continued)

Cranial nerve examination revealed that there was fine nystagmus bilaterally on lateral gaze. Cranial nerves were normal except that she had grossly dysarthric speech. She had mild dysmetria on finger to nose, and diffuse weakness in all extremities (4/5). Her DTRs were 3 to 4+ throughout and there was ankle clonus bilaterally. Her gait was markedly ataxic, and tandem gait could not be performed.

MRI of the brain revealed multiple white matter lesions, many of which enhanced with gadolinium.

Because of the aggressive nature of her MS, she was admitted to University Hospital for intravenous immunosuppressives. Over time she improved with continued treatment, but remained unable to walk a straight line, continued to have memory and personality deficits, and was unable to care for her children or hold a job without assistance.

Patient #2. Salesperson with occasional problems with walking
MC was a 42-year-old salesperson for a chemical company with known MS who came to the MS center for the first time in 2004 with dizziness.

She was well until 1984 when she lived in another state and developed left hemiparesis and was diagnosed as having MS. She had a number of attacks in the first 3 years after diagnosis, including partial blindness in one eye, bilateral leg weakness transiently requiring a cane, and a hemisensory syndrome. Each of these attacks resolved within a few weeks without residua. From 1987 to 2004, she had had no attacks and had no symptoms of her MS.

She developed "dizziness" about 2 weeks prior to her 2004 MS center visit, consisting of some light-headedness and difficulty with walking. This resolved after a few days.

Her neurological examination in 2004 was completely normal, but MRI of the brain revealed multiple white matter lesions, of which two were enhancing. The patient refused medication and has been followed clinically and radiologically over the last 6 years and continues to have no new neurological symptoms and no neurological disability. One MRI scan in the interval between 2004 and 2010 showed two enhancing lesions but this was not associated with symptoms. She has continued to not wish any medication for her MS.

Author's note. These two patients, both with the diagnosis of MS, have very different histories. In the first patient her MS was devastating and led to a marked permanent and progressive worsening of her quality of life. In the second patient her MS, although quite active early in her disease, was only a minor problem for more than 20 years, and she remains without any disability after 26 years of the disease.

7.1   "Form of MS"

In the early twentieth century, MS was classified by the location of the major involvement: brain stem or spinal cord forms. For several decades, neurologists have classified MS according to the occurrence of relapses, as described above. Thus, the most common form of MS is RRMS. Since the efficacy of current therapies are predominantly in decreasing relapses, PPMS patients have not been entered into most recent studies, and the indications for commonly used therapies, do not include patients with PPMS.

Some neurologists use the term SPMS to classify patients with MS later in the disease when relapses usually begin to be less frequent, but disability progression continues. The term is misleading because there is almost always progression in the CNS damage in the disease, even in early stages of the disease. Also, the term "secondary" implies that the disability is secondary to the relapses early in the disease and there is

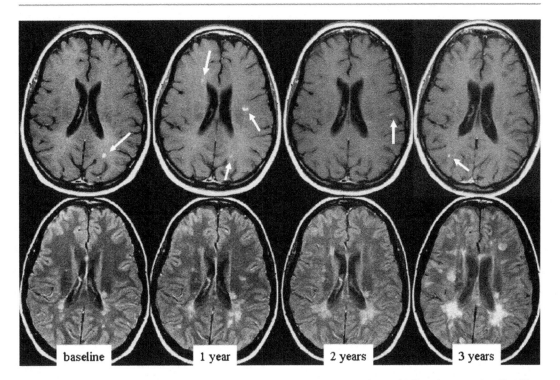

baseline

1 year

2 years

3 years

Fig. 4.2 MRI of the brain: Serial images over time with new lesions appearing and disappearing. Serial gadolinium-enhanced T1-weighted (*upper row*) and FLAIR (*lower row*) MRI images of the brain in a patient with a typical relapsing form of MS and progressive disability. Note the new lesions that appear during this 3-year follow-up, some of them showing gadolinium enhancement (*arrows*)

substantial evidence that this link, i.e., of disability with relapses, does not indeed exist.

7.2 Disability

Neurologists also classify patients by disability. The most commonly used disability scale is the EDSS described above, also called the "Kurtzke scale" after its originator, John Kurtzke. On this scale, the degree of abnormality of a variety of functions, quantitated on "functional scales," are combined with measures of gait. On this scale, 0 is normal, i.e., no disability, and 10 is death from severe MS.

Another recently developed disability scale is the "multiple sclerosis functional composite score" (MSFC). The three components of this scale are the 9-hole peg test, which measures hand strength and dexterity; 25 foot timed walk, a gait measure; and the paced auditory serial addition test (PASAT), a test of cognitive function.

Although this measure of disability is not as broadly used as the EDSS, and is relatively new, it has significant advantages, including testing of cognitive function, a frequent but underassessed problem in MS, and the ability to be used by health care professions other than neurologists.

7.3 Activity

Sometimes MS is classified according to its activity as measured by attack frequency or number of acute lesions on MRI. Many of the drugs currently used in MS are thought to function by decreasing the activity of the disease. The attack frequency correlates fairly well with MRI activity, so that patients with multiple gadolinium-enhancing lesions at any point in time tend to have frequent attacks; gadolinium is a compound frequently injected intravenously during MRI scans to detect active MS lesions and is described in Chap. 5 (Fig. 4.2). This has led some MS investigators to propose MRI as the primary endpoint measure for clinical trials [14], while others consider this to be a poor idea [16].

7.4   Severity

Although relapses are disturbing, most MS patients and neurologists would use progression of disability as their main determinant for severity. That is, they would classify a patient's MS as being severe if disability is being rapidly accrued. In 2005, a large group of European neurologists published results of a study of 9,892 patients which classified their MS by rapidity of accrual of disability on a decile scale, where 1 (e.g., progression to an EDSS of 2 after 19 years of the disease) was very mild MS and 9 very severe MS (e.g., progression to an EDSS of 6, i.e., needed an aid to ambulate, after only 4 years of disease). The ideal therapy should ameliorate the severity of a patient's MS over the long term, but no currently available therapy has been shown to do that [17]. Increasingly, investigators are using this MSSS in clinical research in MS.

Multiple sclerosis and *the Blind Men and the Elephant*: MS remains a mysterious disease, which changes the lives of the patients who have it and the physicians who care for these patients. Because of our lack of knowledge about the etiology, pathogenesis, proper therapy, and our inability to predict the future course of any MS patient, each individual will see the disease and attempt to assess the same disease in different ways. The patient may be particularly concerned about her present disability, trying to maximize day-to-day quality of life and mobility. The patient's family may be focused on the severity, seeing the accrual of disability, and comparing what their loved one can do now compared to what they did a year or two before. Pharmaceutical companies and drug regulatory agencies focus on the attacks since currently therapies are approved or denied based on their ability to decrease attack frequency. Radiologists evaluate the extensive demyelination, try to develop techniques to measure neuronal pathology and atrophy in the disease, and

advocate MRI measures to quantitate the disease. Pathologists study the extensive tissue injury in the CNS, but argue about whether there are different forms of pathology, or a unitary process. The situation is similar to the Indian story of the elephant and the blind men who say the elephant is a rope, a pipe, a hand fan, a branch, a pillar, or a wall depending on what part of the animal they are feeling at the time. These multi-faceted aspects of MS are especially relevant to the issue of therapy (see Chap. 6).

References

1. Waksman BH. Demyelinating disease: evolution of a paradigm. Neurochem Res. 1999;24(4):491–5.
2. Efficacy and toxicity of cyclosporine in chronic progressive multiple sclerosis: a randomized, double-blinded, placebo-controlled clinical trial. The Multiple Sclerosis Study Group. Ann Neurol. 1990;27(6): 591–605.
3. McAlpine D, Compston ND, Lumsden CE. Multiple sclerosis. Edinburgh: E & S Livingston; 1955.
4. Poser CM, Brinar VV. Diagnostic criteria for multiple sclerosis: an historical review. Clin Neurol Neurosurg. 2004;106(3):147–58.
5. Massacesi L, Genain CP, Lee-Parritz D, Letvin NL, Canfield D, Hauser SL. Active and passively induced experimental autoimmune encephalomyelitis in common marmosets: a new model for multiple sclerosis. Ann Neurol. 1995;37(4):519–30.
6. Lucchinetti CF, Bruck W, Rodriguez M, Lassmann H. Distinct patterns of multiple sclerosis pathology indicates heterogeneity on pathogenesis. Brain Pathol. 1996;6(3):259–74.
7. Barnett MH, Parratt JD, Pollard JD, Prineas JW. MS: is it one disease? Int MS J. 2009;16(2):57–65.
8. Ebers GC. Environmental factors and multiple sclerosis. Lancet Neurol. 2008;7(3):268–77.
9. Ramagopalan SV, Knight JC, Ebers GC. Multiple sclerosis and the major histocompatibility complex. Curr Opin Neurol. 2009;22(3):219–25.
10. Cadavid D, Wolansky LJ, Skurnick J, et al. Efficacy of treatment of MS with IFN{beta}-1b or glatiramer acetate by monthly brain MRI in the BECOME study. Neurology. 2009;72:1976–83.
11. Trapp BD, Nave KA. Multiple sclerosis: an immune or neurodegenerative disorder? Annu Rev Neurosci. 2008;31:247–69.
12. Kurtzke JF. Rating neurologic impairment in multiple sclerosis: an expanded disability status scale (EDSS). Neurology. 1983;33(11):1444–52.
13. Roxburgh RH, Seaman SR, Masterman T, et al. Multiple Sclerosis Severity Score: using disability

and disease duration to rate disease severity. Neurology. 2005;64(7):1144–51.

14. Lincoln JA, Cadavid D, Pollard J, et al. We should use magnetic resonance imaging to classify and monitor the course of multiple sclerosis. Arch Neurol. 2009;66(3):412–4.

15. Furby J, Hayton T, Anderson V, et al. Magnetic resonance imaging measures of brain and spinal cord atrophy correlate with clinical impairment in secondary progressive multiple sclerosis. Mult Scler. 2008; 14(8):1068–75.

16. Daumer M, Neuhaus A, Morrissey S, Hintzen R, Ebers GC. MRI as an outcome in multiple sclerosis clinical trials. Neurology. 2009;72(8):705–11.

17. Daumer M, Neuhaus A, Herbert J, Ebers G. Prognosis of the individual course of disease: the elements of time, heterogeneity and precision. J Neurol Sci. 2009;287 Suppl 1:S50–5.

In the usual forms of multiple sclerosis, i.e. in those with a relapsing and remitting course and evidence of disseminated lesions in the CNS, the diagnosis of multiple sclerosis is rarely in doubt.

Raymond D. Adams and Maurice Victor, *Principles of Neurology*

There is no specific diagnostic test for MS. The diagnosis is usually made on the basis of a characteristic constellation of symptoms, signs, and laboratory findings. There are no diagnostic criteria that are both universally accepted and that neurologists invariably utilize in their day-to-day practice. Given the complexity of the central nervous system, it is not surprising that many problems arise when the MS patient presents with symptoms early in the course of the disease. There are frequent false-positive and false-negative findings, even by neurologists experienced in MS. The difficulties in diagnosis lead to anxiety among patients and their families during the evaluation period. The careful neurologist always makes the diagnosis of MS provisionally even in apparently obvious cases, assessing the odds over time, with the probability of MS always being less than 100%. Complete certainty is neither possible nor necessary to make the diagnosis of "definite MS."

If MS is so difficult to diagnose in its early stages, how can Adams and Victor write, "the diagnosis is rarely in doubt?" It is because the diagnosis becomes clearer and clearer over the course of time. Within the first few years of presentation to the neurologist, most patients with MS declare themselves with evidence of new disease activity consistent with the diagnosis, and what might have been an 85 or 95% probability of the diagnosis at the first evaluation, increases to 99% or higher.

As noted in the previous chapter, the histopathologic hallmark of MS is a combination of inflammation, demyelination, and axonal injury limited to the CNS in the absence of any possible alternative diagnosis. Substantial CNS activity early in MS usually does not translate into severe symptoms. Patients with MS early in their disease prior to seeing a neurologist frequently have symptoms that appear and then disappear after days to weeks. These symptoms can be nonspecific and are frequently not helpful in differentiating MS from other diseases. Some symptoms are very common, such as decreased visual acuity, double vision, numbness and tingling, and problems with balance; others such as problems with hearing are less common.

Ultimately, a major symptom develops, which may or may not be more severe than previous episodes, but brings the MS patient to a physician. In the absence of any other obvious disease

A.R. Pachner, *A Primer of Neuroimmunological Disease*,
DOI 10.1007/978-1-4614-2188-7_5, © Springer Science+Business Media, LLC 2012

process, when early MS is suspected, the first episode, the first highly symptomatic relapse or attack, is also labeled "clinically isolated syndrome (CIS)," because it is clinically isolated in time. Many patients have no history of previous symptoms and their CIS is their first manifestation of neurological disease. Other patients have had vague, transient symptoms for which they did not seek medical attention. For yet others, some symptoms had occurred in the past, but had been not been attributed to MS, and the symptoms had resolved. It is at this point that the neurologist is faced with the challenge of making a diagnosis of CIS versus some other neurological disease process.

It must be stressed that CIS is a diagnosis that prejudices physicians toward MS and is considered by most neurologists to be a "pre-MS" condition. The diagnosis of CIS thus should only be made in the absence of an alternative diagnosis. Although CIS or MS can certainly coexist with other diseases, most neurologists will use "Ockham's razor" (see Inset 5.1) in their approach to a relatively uncomplicated new patient, and will want to make a single diagnosis. Similar to MS being a less than 100% certain diagnosis, the diagnosis of CIS must always be made with the continued realization that other diagnoses remain possible.

Inset 5.1 Ockham's Razor

In Latin, this maxim is *pluralitas non est ponenda sine necessitate* or "plurality should not be posited without necessity." Attributed to the fourteenth century Franciscan friar and logician, William of Ockham, the application to medicine is that of "diagnostic parsimony," i.e., that a doctor should try to look for the fewest possible causes, preferably just one, to explain a patient's symptoms. Ockham's razor in medical diagnostics is an example of a heuristic, an intuitive approach or educated guess or common sense rather than a formal and rigid approach to a problem.

---

1   History and Examination

MS almost always has its onset in young people 20–35 years old. The mean age of patients entered into studies of multiple sclerosis is usually in the early to mid-30s. Preadolescent or middle-aged individuals are less likely to develop the disease than people in their 20s and 30s. Although more women have MS than men, the 2:1 ratio is not particularly helpful diagnostically. Even though most patients do not have a blood relative with MS, a positive family history increases the likelihood of MS.

A careful history and thorough general and neurological examination are critical for the patient with a possible diagnosis of CIS to identify conditions causing disease outside of the CNS, and thus to "rule out" CIS, and to identify findings on neurological examination consistent with CNS injury, i.e., to "rule in" CIS. A history consistent with any of the possible "MS mimickers" (Chap. 6) or underlying medical conditions that predispose to cerebrovascular disease, such as hypertension, diabetes, smoking, contraceptive use, or positive family history needs to be sought. Risk factors for HIV infection, such as intravenous drug abuse, promiscuous sexual practices, or HIV in a sexual partner need to be ascertained. Other medical conditions which can lead to neurological symptoms such as systemic lupus erythematosus (SLE) need to be considered. In areas endemic for Lyme disease, a history of an unusual skin rash, facial paralysis, or chronic headache syndrome raises the possibility of *Borrelia burgdorferi* infection. Enlarged lymph nodes raise the possibility of lymphoma or sarcoidosis.

Most patients with CIS will have an abnormality on neurological examination consistent with their symptoms. Thus, patients complaining of decreased vision who have optic neuritis will have decreased visual acuity in the affected eye and abnormal pupillary responses. Patients with diplopia will have abnormal eye movements. Patients who have complaints of weakness or numbness caused by spinal cord involvement (myelitis) will have diminished strength, sensation, and increased deep tendon reflexes in their arms and legs. In addition, some patients with CIS will have evidence of more than one area of involvement of the CNS. For instance, a patient with new symptoms of leg weakness referable to a spinal cord lesion might have abnormal pupils on examination consistent with previous, but asymptomatic, involvement of an optic nerve.

2 Laboratory Findings

2.1 Routine Studies: Blood, Urine, Chest X-Ray

MS is limited to the CNS, so blood analysis should be completely normal. Standard blood tests (electrolytes, BUN, glucose, metabolic panel including liver function tests, complete blood count with differential) are a necessary part of the initial diagnostic evaluation. Abnormalities such as anemia, an elevated white count, or abnormal liver function tests should prompt an evaluation that might identify an alternative disease process. Occasionally, a urinary tract infection (UTI) triggers an attack. If an UTI is not present, urine testing should be normal. Elevated protein or glucose in the urine or blood would be of concern for another disease process. A chest X-ray can pick up a number of mimics of MS (e.g., neurosarcoidosis, metastatic cancer, or infections such as tuberculosis or fungal infections).

2.2 CNS Imaging

The test which is the most helpful, but is also most frequently misinterpreted, is the MRI scan of the brain. The finding of white matter lesions (white matter lesions) on an MRI of the brain or spinal cord is highly sensitive, but not specific for the diagnosis of MS (see Inset 5.2). Thus, a sizeable majority of patients who ultimately develop MS will have white matter lesions on brain MRI, but the presence of some white matter lesions, especially in a middle-aged population is not rare in individuals who do NOT have MS. White matter lesions which have no apparent pathological significance are sometimes referred to as "UBOs"

(unidentified bright objects). It is sometimes difficult to determine whether unexpected areas of brightness on the MRI in individuals over the age of 40 represent worrisome areas or benign UBOs. White matter lesions are nonspecific and can be seen in individuals with hypertension, migraine, and a variety of other conditions (Inset 5.2).

Inset 5.2 A 34-Year-Old Construction Worker with Leg Weakness Incorrectly Diagnosed as MS

JH was a 34-year-old male construction worker who began to develop difficulty with decreased strength in his legs at work a few weeks prior to admission to University Hospital. As a construction worker whose job entailed heavy manual labor, he needed full strength in his legs. His weakness progressed and he came to the ER of a local hospital because he could no longer work. On examination in the ER he had weakness in the legs and increased reflexes. Because of an abnormal spinal cord MRI scan, with a cervical white matter lesion, he was given a diagnosis of multiple sclerosis, placed on corticosteroids and discharged home after 5 days of treatment. However, his weakness progressed and he was admitted to the University Hospital ER. History revealed that he had been previously healthy. He had come to the USA from Ecuador and had been living in the USA for 2 years. On examination, he had marked weakness in his legs, primarily in the iliopsoas and quadriceps and had increased deep tendon reflexes and a Babinski sign. Imaging of his cervical and thoracic spine revealed enhancing lesions, but most were felt to be extramedullary. MRI of his brain was consistent with neurocysticercosis. Surgical exploration revealed multiple extradural masses which were cysts of the racemose form of neurocysticercosis. He was treated with surgical excision of the cysts, and on treatment with albendazole and corticosteroids, his clinical condition slowly improved.

(continued)

Inset 5.2 (continued)

Author's note. This patient's story accentuates the necessity of being particularly cautious in making a diagnosis of MS in a patient with involvement of only one clinically defined area of the central nervous system. This patient's presenting symptoms were solely referable to the cervical spinal cord. The case also demonstrates the fact that white matter lesions are seen in many other diseases besides MS.

Inset 5.3 A 38-Year-Old Woman with a Possible Diagnosis of MS

A 38-year-old woman came to the MS Center at University Hospital with symptoms of left leg weakness and the question of diagnosis of MS. Four years previously, she suddenly developed decreased vision in the right eye associated with eye pain. This had been evaluated and found to be optic neuritis which resolved spontaneously. Two months prior to her presentation at the MS Center, she had developed left leg weakness with numbness and tingling in both her left and right legs. Her primary care physician referred her to a neurologist who suspected MS and obtained an MRI of the brain, which was read as normal. However, the brain MRI was obtained on an "open" MRI, since the patient was claustrophobic, and "open" MRIs are known to be less sensitive. The neurologist was loathe to make the diagnosis of MS with the normal MRI, and scheduled the patient for a 6-month follow-up visit. The primary care physician then referred her to our center.

On examination at our center she had mild optic atrophy and decreased visual acuity in the right eye, 4/5 weakness proximally in the left leg, hyperreflexia and clonus in the left leg and a left Babinski

reflex. Our review of her previous MRI of the brain revealed it to be suboptimal technically. The brain MRI was repeated, this time on a closed MRI, showing a few periventricular white matter lesions. An MRI of the cervical spine revealed two parenchymal white matter lesions, a lumbar puncture revealed oligoclonal bands, and the anti-aquaporin antibody test was negative, making the diagnosis of neuromyelitis optica less likely. We felt that she had definite MS, and she was begun on interferon-β therapy. Subsequent follow-up in the years following the initial evaluation confirmed the diagnosis of MS.

Author's note. By having two or more attacks with objective clinical evidence of two or more lesions, the patient fulfills McDonald's criteria for MS, and the initial neurologist did not have to confirm the diagnosis of MS with an MRI of the brain. We performed extra testing to confirm the diagnosis, and each test was consistent with the diagnosis of MS.

This patient's story demonstrates two lessons. First, that all patients with MS do not have to have MRI lesions in the brain at onset of their disease, i.e., the clinical history and findings on exam can be sufficient to make the diagnosis. Second, the quality of MRIs of the brain are important in looking for white matter lesions, and suboptimal scans, i.e., inadequately sensitive scans, can be potentially counterproductive in the evaluation of patients with possible MS.

The presence of white matter lesions on MRI is by far the most significant finding in many patients referred to neurologists for possible MS. The MRI appearance of white matter lesions depends on the peculiarities of myelin, water molecules, and magnetic fields. Myelin is a tissue with a very high lipid content and a low water content. When myelin is lost, it is replaced by tissues that have a relatively high water content. Since such a large part of the MRI is based on water molecules, replacement of the very low water signal in normal myelin by the increased water signal in damaged myelin is striking. Abnormal characteristics of brain tissue in patients with MS other than increased water signal, such as axonal injury, inflammation, and atrophy are difficult, if not impossible to visualize on routine MRI.

There are no specific brain MRI patterns which are 100% diagnostic of MS, but some characteristics of white matter lesions are very helpful diagnostically. Lesions which are adjacent to the ventricles of the brain, especially those which are aligned perpendicularly to the ventricles and are flame-shaped, sometimes called "Dawson's fingers," are particularly suggestive (Fig. 5.1). Although most large lesions occur in the deep white matter of the brain, smaller juxtacortical lesions, involving white matter connecting areas of cortical gray matter, are increasingly being recognized. These may be increasingly identified as higher resolution MRI scanners using larger magnets come into wider use. These lesions may be especially relevant to the progression of disability, such as cognitive dysfunction. The power of a magnetic field is measured in a unit called a Tesla after Nikola Tesla, who was a Serbian inventor and electrical engineer, responsible for the discovery or development of a number of advances including alternating current, electromagnetic fields, wireless communication, and robotics. The MRIs most commonly used now by neurologists have magnetic fields of 0.5–1.5 T, but many centers have 3 T magnets, and a limited number of MRI facilities, mostly in research settings, currently use 7 or 8 T magnets.

Frequently, neurologists concerned with the possibility of MS will also request that the MRI of the brain be supplemented with an MRI sequence using an intravenous contrast agent, standardly a chelated gadolinium compound. With gadolinium-enhanced MRI, a recent MS lesion may demonstrate enhancement since the compound will leak out into the lesion in the presence of disrupted blood–brain barrier around areas of active inflammation. On MRI scanning the leaked gadolinium will appear as one or more circular or spherical areas of brightness; this can be seen on the serial

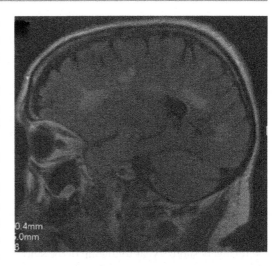

Fig. 5.1 Dawson's finger. James Walker Dawson (1870–1927) was a Scottish physician whose work provided an invaluable link between Charcot's studies in the late nineteenth century and those in the twentieth century. He systematically characterized the pathology of MS lesions and described the tendency of lesions to form along large periventricular veins. These lesions, whose long axis tends to be perpendicular to the ventricules when seen on MRI, on the sagittal section shown in this figure, are called "Dawson's fingers," in tribute to his major contributions to the field

MRIs in a patient over time depicted in Fig. 4.2. Over the course of weeks to months after the onset of a new lesion, the blood–brain barrier will return to normal and the lesion will no longer be "gadolinium-enhancing." Some patients with MS will frequently have gadolinium-enhancing lesions on their brain MRIs and others only rarely so. Thus, the absence of a gadolinium-enhancing lesion does not contradict a diagnosis of MS. However, the presence of one or more such lesions does aid in the diagnosis. Gadolinium enhancement is not specific for MS and can occur in a variety of diseases. However, the combination of typical "Dawson's fingers" on MRI along with gadolinium enhancement of one or more white matter lesions is highly suspicious for MS.

Of all the tests involved in attempting to ascertain the diagnosis in a patient with a neurological event looking like CIS, the MRI of the brain is arguably the most important, because of its high predictive value for the development of MS. In a patient presenting with a clinical syndrome indicating inflammation of the optic nerve, brainstem, or spinal cord and periventricular white matter lesions in the brain typical of MS, the chance of developing MS is extremely high. Thus, in the first study of treating CIS with an MS disease-modifying drug, the CHAMPS study [1], an inclusion criterion was the presence of two or more clinically silent lesions of the brain at least 3 mm in diameter on MRI scans and characteristic of multiple sclerosis with at least one lesion being periventricular or having an ovoid shape. This criterion was included because the investigators wanted a group with a high likelihood of developing definite MS, and indeed in the relatively short 3-year follow-up of the study, 50% of patients in the placebo-treated group developed definite MS, a number which would have likely been lower had the inclusion criteria not included the brain MRI lesion requirement.

2.3   Cerebrospinal Fluid Analysis

The most frequent abnormality in the CSF of MS patients is the presence of oligoclonal immunoglobulin bands (OCBs) in the CSF not present in the serum. This finding occurs in more than 95% of MS patients [2] when optimum methodologies for OCBs are used. OCBs consist of immunoglobulin G, similar in size and charge, which travel together on isoelectric focusing gel electrophoresis (Fig. 5.2). Although OCBs are not specific for MS and occur in neurological infections such as neurosyphilis, in the proper clinical setting their presence almost always confirms a diagnosis of MS in a patient with a borderline clinical presentation [3]. Since the blood–brain barrier usually prevents significant amounts of serum IgG from entering the brain and CSF, CSF OCBs are products of IgG production within the CNS. Plasma cells, the terminally differentiated B cells which produce immunoglobulin, are a prominent part of many MS lesions, and nests of plasma cells have occasionally been identified in the meninges of MS patients. Plasma cells in the CNS and OCBs in the CSF in diseases other than MS are a feature of chronic CNS infections, suggesting that MS may

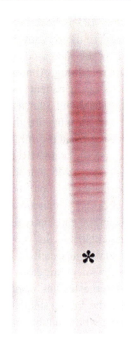

Fig. 5.2 Oligoclonal bands (OCBs) are related populations of immunoglobulins in the CSF (*right lane*, below) that travel together on a special type of gel called IEF (isoelectric focusing) and indicate the production of IgG within the CNS. The *lane* to the *left* represents a similar treatment of serum from an MS patient showing the absence of OCBs

indeed be a chronic infection. The precise basis for the prominent immunoglobulin production within the CNS remains unexplained, but recent evidence points to the proclivity of the CNS to support antibody production and downregulate cellular immune reactions.

MS is an inflammatory disease with mononuclear cells, lymphocytes and monocytes, infiltrating the brain. This infiltration is mirrored frequently by a modest increase in lymphocytes in the CSF, often associated with a slightly elevated protein level Although elevated mononuclear cell counts and protein are common in MS, they are not so invariable that their absence argues against the diagnosis. In fact, one of the mysteries of MS is that the brain parenchyma can be highly inflamed while mononuclear cells and normal protein are absent or only marginally elevated in the CSF. High CSF mononuclear cell counts greater than 100 in the CSF strongly suggest an alternative diagnosis.

This disconnection between parenchymal inflammation and inflammatory cells in the CNS indicates that in this disease the CSF is not invariably a good measure of the severity of CNS inflammation.

Patients in whom the diagnosis of MS can be made at the first visit to the neurologist. Some patients present to the neurologist not with their first attack but with their second or third. Their earlier neurological episodes may have been ignored or may have been misdiagnosed. Evidence of accumulated damage to the CNS and previous attacks in such patients can aid in making the diagnosis of MS.

Criteria for definite MS. The first commonly accepted diagnostic criteria, the Schumacher criteria, developed in 1965, were purely clinical and stressed the importance of dissemination in space and time. They can be summarized in one sentence: two or more lesions demonstrable in the white matter of the central nervous system separated in time (by at least 6 months) and space, i.e., location within the CNS, with objective abnormalities on the neurological examination, without an alternative diagnosis. Subsequently, the Poser criteria incorporated laboratory testing, creating categories of "laboratory-supported" definite or probable MS. More recently, the McDonald criteria for the diagnosis of definite MS [4] (http://www.mult-sclerosis.org/DiagnosticCriteria.html) have been utilized both in clinical trials and in practice. These are an advance in that they include MRI data. An instance of the usefulness of the McDonald criteria is the patient who has CIS, with one attack and objective clinical evidence of only that one lesion but with multiple MRI lesions ("dissemination in space") and the development of new MRI lesions over time ("dissemination in time"). This patient qualifies as having the diagnosis of definite MS by McDonald criteria months or years before the second attack. Some MS neurologists mistrust the McDonald criteria [5], and for community neurologists the emphasis on MRI can be difficult, but for many neurologists they have become the standard for diagnosis.

References

1. Jacobs LD, Beck RW, Simon JH, et al. Intramuscular interferon beta-1a therapy initiated during a first demyelinating event in multiple sclerosis. CHAMPS Study Group. N Engl J Med. 2000;343(13):898–904.
2. Link H, Huang YM. Oligoclonal bands in multiple sclerosis cerebrospinal fluid: an update on methodology and clinical usefulness. J Neuroimmunol. 2006;180(1–2):17–28.
3. Thompson EJ, Freedman MS. Cerebrospinal fluid analysis in the diagnosis of multiple sclerosis. Adv Neurol. 2006;98:147–60.
4. Polman CH, Reingold SC, Edan G, et al. Diagnostic criteria for multiple sclerosis: 2005 revisions to the "McDonald Criteria". Ann Neurol. 2005;58(6):840–6.
5. Poser CM, Brinar VV. Diagnostic criteria for multiple sclerosis: an historical review. Clin Neurol Neurosurg. 2004;106(3):147–58.

Although the course of the disease in all MS patients is different, most MS patients present in fairly straightforward manner, so that the diagnosis is usually made soon after first presentation to a neurologist. However, sometimes the diagnosis is incorrect, either because the patient has another disease process and MS is mistakenly diagnosed, or less commonly because the patient has MS and another diagnosis is erroneously made. This chapter discusses the diseases that most commonly mimic MS. The confusion usually occurs at the time of the first neurological event, clinically isolated syndrome (CIS), rather than in patients who have had previous events. CIS, as previously discussed in Chaps. 4 and 5, is the sudden development of a demyelinating and inflammatory event, consistent with a first attack of MS, that does not meet criteria for clinically definite MS. Thus, optic neuritis or myelitis or a hemisensory syndrome in a young, previously healthy individual might be CIS or one of the other diseases outlined in the chapter. If the patient with CIS develops new MRI lesions or has new attacks, the diagnosis of MS is much less difficult. Some issues and difficulties involved in differential diagnosis of the CIS patient are highlighted in a recent clinical review of six German patients with CIS [2]. Some examples of diseases that have clinical, imaging, or CSF characteristics similar to MS are summarized in Table 6.1. Finally, not all symptoms localizing to the CNS in patients with known MS should automatically be assumed to be a new attack of MS, as exemplified by the patient in Inset 6.1.

Inset 6.1 A Patient with Known MS and a New Neurological Event—Most Likely MS, But…

KW, a 34-year-old emergency room physician with an 8-year history of MS, presented to her neurologist for emergency evaluation because of weakness and numbness in her arms. Her exam revealed weakness and sensory changes in the arms; there was also decreased strength in her legs. Since in previous attacks she had cervical spinal cord involvement, it was assumed that she was having a spinal cord attack and she was treated with intravenous corticosteroids. She improved but not completely. She continued to be symptomatic and had difficulty at work with suturing and other tasks involving fine motor control in her hands. An MRI scan of the cervical spine showed both an acute MS plaque at C3-5, and, in addition a large central cervical disc herniation at C5 causing cord compression. She underwent decompressive surgery for the disc and had complete recovery of her strength and sensation in her arms and legs.

Author's note. Most acute neurological events localizing to the CNS which occur in patients with MS are rightly assumed to be MS attacks. However, patients with MS are not immune from other disease processes. In this patient, two processes, an

(continued)

A.R. Pachner, *A Primer of Neuroimmunological Disease*,
DOI 10.1007/978-1-4614-2188-7_6, © Springer Science+Business Media, LLC 2012

Table 6.1 Similarities of MS to some other common diseases

Disease	Prevalence	Clinical characteristics	MRI brain characteristics	CSF oligoclonal bands	Relapsing–remitting
Stroke	Common	++	+	–	–
ADEM	Rare	++	++	+	–
Neurosarcoidosis	Rare	+	+	+	+
Lyme neuroborreliosis	Common (in endemic areas)	+	–	+	–
HIV/OI	Common	+	+	+	–
SLE/APS	Rare	+	+	–	+
Migraine	Common	–	+	–	–
Lymphoma	Rare	+	+	+	+
Systemic vasculitis	Rare	+	+	+	+

ADEM acute disseminated encephalomyelitis; *HIV/OI* human immunodeficiency virus or its opportunistic infections; *SLE/APS* systemic lupus erythematosus/anti-phospholipid antibody syndrome

Inset 6.1 (continued)

active MS plaque in the cervical cord and a cervical disc, were present which together resulted in a very symptomatic patient. Her symptoms only resolved after both processes were addressed.

1 Cerebrovascular Disease/Stroke

In any community within the USA or Europe, the most common cause of rapid onset of a neurological deficit, combined with focal lesions on MRI, is not MS, but cerebrovascular disease. Thus, it is not surprising that patients with stroke are sometimes diagnosed as having MS and vice versa. This is increasingly true now in the USA for two reasons: first, the MRI is being used more and more in evaluating neurological symptoms, and second, it can be difficult to distinguish white matter lesions in the brain MRI caused by cerebrovascular disease from those caused by MS. The population in whom this is usually a problem are patients between 40 and 60 years old, for whom underlying precipitating factors such as hypertension, diabetes, hyperlipidemia, and smoking may have predisposed to atherosclerosis, but the patient is still young enough that multiple sclerosis must be considered. Alternatively, in a young patient without obvious predisposing

factors stroke mimicking MS can occur from unusual causes of stroke, such as the patient with an atrial myxoma described in Inset 6.2.

Inset 6.2 A 50-Year-Old Woman with Double Vision: Stroke Mimicking MS

BE was a 50-year-old woman with no previous medical history when she presented to a hospital emergency room with double vision. Six months previously, she had visited her primary care doctor for an episode of severe vertigo and nausea which began to improve spontaneously after 4 days and resolved after 2 weeks. Her examination was consistent with a very mild internuclear ophthalmoplegia (INO) and she had no other findings on examination. On the basis of multiple brain MRI white matter lesions and no alternative diagnosis, she was diagnosed with multiple sclerosis, despite negative oligoclonal bands in the CSF. She was offered MS therapy, which she decided not to take.

She was subsequently evaluated by two neurologists who felt that she had probable multiple sclerosis.

Seven months after her episode of double vision, she was found to have an absent radial pulse in her left arm which turned out to be due to a clot in the radial artery. Investigation for a source of the clot

(continued)

revealed a mass in her heart, which proved to be an atrial myxoma, a benign cardiac tumor which can send emboli to the brain. She underwent surgery for removal of the myxoma, and had no further symptoms. In retrospect, the white matter lesions in her brain represented sites of embolic infarction not MS lesions, and the episodes of neurological symptoms were ischemic from emboli not MS attacks.

Author's note. This patient had a pattern resembling MS when she presented with double vision, despite turning out to have embolic strokes caused by the tumor in her heart. At presentation, her age was a bit high for the typical initial diagnosis of MS but not inordinately so. However, at that age, and with a negative spinal tap, the diagnosis should have been in question.

Another reason that stroke can be confused with MS is that stroke can satisfy Schumacher's criteria for MS, i.e., at least two separate CNS events disseminated in time and space. Schumacher's criteria also include the fact that there must be no better explanation for the neurological picture, but sometimes the diagnosis of stroke is not considered a "better" explanation. An example of how a young woman was mistakenly diagnosed as having strokes and actually had MS is provided in Inset 6.3. Thus, for the reasons outlined above, in addition to the fact that it is so common, stroke is the most frequent mimic of MS. A few distinguishing features are listed below:

1. MS patients are usually young individuals who are healthy other than their MS, while stroke patients usually are older and have risk factors for stroke, such as hypertension, diabetes mellitus, smoking, and hyperlipidemia (high levels of lipids in their blood).
2. MS patients have oligoclonal Ig G bands in their CSF >90% of the time, while stroke patients almost never have them.
3. On MRI active MS lesions usually do not have increased signal on diffusion-weighted images, while strokes usually do.

4. MS patients frequently have evidence by MRI of lesions in their spinal cords, while stroke patients almost never do.
5. Acute MS lesions often enhance after gadolinium injection, while enhancement is less commonly seen in stroke lesions.

In other patients who come to the attention of their physicians for nonspecific sensory symptoms

Inset 6.3 A 37-Year-Old Woman with Acute Severe Left-Sided Weakness—MS Mimicking Stroke

SB was a 37-year-old woman who presented to the hospital with a left hemiparesis. She had a history of a prior stroke causing right hemiparesis 3 years previously, which had resolved to a great extent and she was ambulatory with a cane prior to admission. Because of the history of the previous stroke and an elevated blood pressure on admission, the diagnosis of a right hemispheric stroke was made, and she was admitted to the hospital's stroke ward.

After the first few days, the patient did not improve and an MRI scan of the brain was performed with diffusion studies and gadolinium injection. There was no area of restricted diffusion, which would have been consistent with an acute stroke, but there was gadolinium enhancement of a number of lesions, consistent with active MS plaques. Cervical cord MRI and triple evoked responses and CSF analysis were consistent with MS, and CT angiogram, showed no cerebrovascular disease. A lumbar puncture was performed and the CSF revealed positive oligoclonal bands; a diagnosis of MS was made. It was felt in retrospect that the "stroke" at age 34 was her first attack of MS.

Author's note. The physicians seeing this patient in the ER made a mistake involving the "anchoring heuristic," in which the previous diagnosis of stroke was excessively relied upon, without factoring in new information [16]. In diseases causing focal neurological lesions in young woman, MS must be considered very high on the list.

or headache, an MRI is frequently obtained, white matter lesions are seen, and the patient is referred to a neurologist. The most common diseases in which this occurs is migraine and hypertension, and a competent history and examination will generally quickly identify those patients.

2   Neurological Infections

2.1   HIV

Infection with HIV is often subclinical for an extended time period, and individuals with HIV frequently are immunodeficient without symptoms until they get an opportunistic infection (OI). The likelihood that OIs will get confused with MS is highly variable depending on geographic region. Generally, HIV OIs are infrequently confused with MS, so the problem only becomes a significant one if the prevalence of HIV in a community is high. For example, in Newark, NJ, the prevalence of HIV/AIDS in the general population is reported to be 2–3%, so the likelihood for an OI and MS being confused is higher than other areas where the HIV/AIDS prevalence is lower. The most common OIs which cause focal inflammatory CNS disease mimicking MS are toxoplasmosis and progressive multiple leukoencephalopathy (PML). Another disease affecting immunosuppressed individuals with HIV and which can mimic MS is primary CNS lymphoma, which some investigators feel is caused by Epstein–Barr virus (EBV). Toxoplasmosis and primary CNS lymphoma generally have characteristic imaging appearances that do not look like MS, but PML can cause white matter disease that can look very similar to MS. Neurological involvement in HIV and its OIs and primary CNS lymphoma are discussed in later chapters in this book.

2.2   Neurosyphilis

Lord Brain's seventh edition of his textbook on neurology includes meningovascular syphilis and tabes as high on the list in the differential diagnosis of MS, possibly because of the fact that neurosyphilis was more common at that time [3]. In fact, from the time effective syphilis treatments became available at the turn of the twentieth century until World War II and the availability of penicillin, syphilis treatments were also used in MS. The rationale was that both diseases caused significant disability and, since there was no treatment for MS, it was worth trying antisyphilitic therapies [4]. Neurosyphilis is less common now, but still occasionally causes diagnostic problems, as in the patient in Inset 6.4. The diagnosis can be made by positive syphilis

---

Inset 6.4 Meningovascular Syphilis Mimicking MS: A 41-Year-Old Man with Trouble Talking

A 41-year-old man presented to the ER with sudden development of dysarthria (difficulty with motoric articulation of his words), right facial weakness, and left-sided weakness. He had no significant past medical history, including no risk factors for stroke. His examination revealed findings consistent with a right pontine lesion. MRI of the brain revealed a white matter lesion in his right pons without mass effect as well as six white matter lesions scattered throughout the brain; none of the lesions were periventricular and none enhanced with gadolinium. The initial diagnosis of MS was made by the ER staff, with a number of other possible alternative diagnoses, and a lumbar puncture was performed to confirm the diagnosis. There were 79 white cells/cu. mm. and an elevated protein. The CSF VDRL test, a test for syphilis (see Chap. 13) was positive at 1:8 and the serum VDRL was 1:16. Oligoclonal bands were positive. The patient's wife was seropositive for syphilis also. Both were treated with penicillin. In follow-up, patient resolved his findings, the VDRL slowly dropped and finally became negative, and the patient had no further clinical or MRI lesions.

Author's note. Neurosyphilis remains a clinically important problem, with prevalence of the infection increasing in men who have sex with men (MSM), and a masquerader of other diseases. Had this patient's physician not ordered a syphilis test, the diagnosis and proper therapy would have been missed.

tests in the CSF and blood; it must be remembered that sometimes patients with neurosyphilis have negative syphilis tests in the serum but positive tests in the CSF. Neurosyphilis is discussed at length later in Chap. 12.

2.3 Neurocysticercosis

Neurocysticercosis is a chronic CNS infection caused by ingestion of eggs of the tapeworm *Taenia solium*. Pigs and humans are reservoirs of cysticercosis, and the disease in humans occurs in parts of the world in which there is a significant rate of infection in pigs. Most of the neurocysticercosis cases in the USA are from Mexico, as the disease is endemic there in humans and pigs, as it is in many countries in Central and South America. Since pigs in the USA do not harbor *Taenia solium*, neurocysticercosis in the USA will be seen generally in areas with a large immigrant population, and thus clinicians in those areas must be aware of this multifocal CNS disease that can affect a young population and mimic MS. The disease is treatable with the benzimidazole compound, albendazole (see case of neurocysticercosis in Chap. 5); praziquentel is also effective, but costs more.

2.4 Lyme Disease

A chronic infection caused by various strains of the spirochete *Borrelia burgdorferi*, Lyme disease is a relatively common cause of illness in endemic areas in the USA and Europe. There are differences between the American and European variants with serious neurological disease more common in the European forms of the infection [5]. Because the illness is inflammatory and can be limited to the nervous system, in the occasional patient the disease can be mistaken for MS. The diagnosis of Lyme neuroborreliosis (LNB) can be made by the demonstration of high levels of anti-*B. burgdorferi* antibody in the CSF and usually also in the serum in an individual who has had exposure to ticks in an area endemic for Lyme disease. In addition, promi-

nent white matter disease in the brain or spinal cord seen on MRI scans is highly unusual for LNB and thus tends to exclude the diagnosis. Since many individuals in endemic areas can have serum antibody positivity on the basis of exposure, a positive Lyme serology in the blood does not confirm the diagnosis of LNB; thus, patients with MS in Lyme endemic areas can be seropositive on the basis of *B. burgdorferi* exposure without active infection in the CNS. Lyme disease is discussed at length in Chap. 12.

2.5 Tropical Spastic Paraparesis

Spinal cord involvement is common in MS. In patients with spinal cord involvement who come from parts of the world endemic for the retrovirus, human T-lymphotropic virus-1 (HTLV-1), this infection should be considered as a possible cause of myelitis along with MS. Myelitis due to infection by this pathogen is called tropical spastic paraparesis (TSP) or HTLV-1-associated myelopathy (HAM). Along with spinal cord involvement, patients with TSP can have white matter lesions in the brain on MRI, although these are not usually associated with clinical symptoms. The vast majority of individuals in endemic areas with positive anti-HTLV1 antibodies in the blood do not have TSP, so sometimes HTLV-1 antibody positivity can occur in patients with true MS. In such patients, it might be difficult to distinguish MS from TSP, especially if the disease is primarily in the spinal cord. TSP is discussed more fully in Chap. 12.

2.6 Tuberculosis

In parts of the world with a large burden of tuberculosis (TB), such as India, the diagnosis of neurological involvement with Mycobacterium tuberculosis, the TB bacillus, is often made as a default in patients with neuroinflammatory disease and anti-TB therapy is initiated. Thus, patients with MS presenting with a first attack

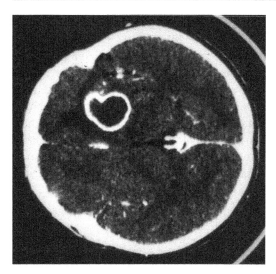

Fig. 6.1 CNS TB lesion. Tuberculosis infection in the brain can appear as a tuberculoma on MRI scanning, although occasionally TB lesions can be confused with MS lesions on brain MRI

can be diagnosed as having TB, anti-TB therapy is initiated, and the diagnosis is "confirmed" by resolution of the neurological symptoms. This scenario is not rare and can delay initiation of appropriate treatment for MS. Although usually CNS tuberculosis presents with different lesions that do not mimic MS (Fig. 6.1), sometimes TB can present as multifocal white matter lesions which can mimic MS on MRI [6]. Tuberculosis is discussed in more depth in Chap. 12.

2.7   Brain Abscess

Since brain abscesses are frequently multifocal and usually present weeks after the initial bacteremia, brain abscesses sometimes present as a new neurological deficit without a known antecedent infection, along with multifocal lesions on MRI mimicking MS. Generally speaking, MRI imaging and the clinical situation can distinguish the two processes, but occasionally patients have required brain biopsy for the diagnosis. Also, patients with very aggressive MS with large lesions, sometimes called tumefactive MS, can be misdiagnosed as having brain abscesses.

3   NMO

Neuromyelitis optica (NMO), also called Devic's disease, is a disease very similar to multiple sclerosis, characterized by demyelination, inflammation, and disability in young individuals. However, the clinically relevant lesions are generally restricted to the optic nerves and spinal cord, in contrast to MS where they occur throughout the CNS, including the brain. On imaging, the lesions in the spinal cord are "longitudinally extensive," meaning that they are longer than a few vertebral levels, in contrast to MS, where they generally span fewer than three vertebral levels. Patients with NMO generally have no or very few lesions on brain MRI, and oligoclonal bands are positive in only 25% of NMO patients in contrast to the 95% positivity rate in MS. For many years, it was considered a forme fruste of MS, but now this view is being challenged, and many neuroimmunologists consider it as being a completely different disease process. It is much less common than MS. In France, the prevalence of MS is estimated at 1 in 600, similar to the prevalence in the USA, while the prevalence of NMO in France has been determined to be 0.2% of that, or approximately 1 case in 300,000 individuals [7]. In Japan, there is a variant of MS called opticospinal MS, which in many patients appears to be identical to the NMO described in Western countries, although it remains unclear whether NMO and opticospinal MS are the same entity. Other features of NMO are discussed in Chaps. 8, 17–19.

There has been a great deal of excitement because of progress over the last decade in elucidating the pathogenesis of NMO. It is characterized by the presence of an autoantibody in the serum, occurring in the majority of NMO patients, which is detected by its ability to bind aquaporin-4 channels in the brains of mice (see Chap. 17 for a description of aquaporins and Chap. 18 on the NMO antibody assay). Models of NMO have been recently induced in rodents by transfer of IgG from the serum of patients with NMO, providing evidence that the autoantibody may be pathogenic and responsible for the clinical phenotype [8, 9].

These findings, as well as a plethora of new para-neoplastic syndromes identified as being antibody-mediated (see Chap. 14), have raised the possibility that at least some of the CNS injury in multiple sclerosis may be autoantibody-mediated, as well as in other neuroimmunological diseases of unknown pathogenesis.

For the practicing neuroimmunologist, it is important to distinguish NMO from MS because the therapies are different for the two diseases. In contrast to MS, where initial therapy is usually interferon-β, glatiramer acetate, or fingolimod, the recommended therapy for NMO is plasmapheresis and immunosuppressive agents, although the effectiveness of these therapies has not been proven. It is generally assumed that recurring attacks in NMO result in progressive injury, in contrast to MS where the link between attacks and progressive disability is less clear. In NMO, immunosuppressive therapy with mycophenolate mofetil has resulted in decreased attacks [10, 11].

4 ADEM

As mentioned in Chap. 3, post-vaccinial encephalomyelitis, one of a group of diseases called acute dissseminated encephalomyelitis (ADEM), was the illness which established the field of neuroimmunology. Usually occurring in children, ADEM can only be reliably separated from acute, highly inflammatory MS by the fact that ADEM is a monophasic event, while MS recurs and progresses. Other differences are that ADEM usually occurs days to weeks after a well-defined acute infection or a vaccination and can cause encephalopathy and seizures, features which usually do not occur in MS. Occasionally, ADEM patients have one or more relapses within a few months after the initial event without developing MS; this presentation is sometimes called multiphasic disseminated encephalomyelitis (MDEM). The clinical characteristics of 28 children with ADEM, 7 with MDEM, and 13 with MS have recently been reviewed [12].

5 Leukodystrophies

The most dramatic and obvious finding in MS, especially on imaging, is demyelination. Thus, it is not surprising that hereditary disorders of white matter, also called leukodystrophies [1], can be misdiagnosed as MS. Usually, neurologists do not consider leukodystrophies as a cause of white matter disease in adults because leukodystrophies are most often overwhelmingly debilitating and rapidly progressive diseases in infants and children. However, recently, there has been increasing appreciation that leukodystrophies in adults are occasionally misdiagnosed as MS [1]. Many adult-onset leukodystrophy patients present with predominantly cognitive symptoms consistent with a subcortical dementia. Late stages are characterized by incontinence, abulia, and increased spasticity. Although adult-onset leukodystrophies do not generally have clinical pictures consistent with relapsing–remitting MS, they can mimic primary progressive MS, in which progressive disability dominates the clinical picture.

6 Other Conditions

6.1 Neurosarcoidosis

Sarcoidosis is an inflammatory disease affecting a wide range of organ systems including the nervous system and is discussed again in Chap. 13 among the systemic inflammatory diseases. Neurosarcoidosis is inflammatory, frequently with white cells and oligoclonal bands in the CSF mimicking MS. Multifocal white matter lesions and cranial nerve palsies, especially optic neuritis, may occur. It is, therefore, sometimes confused with MS. In a recent study of 3,900 patients referred to Vanderbilt's MS center over 13 years, the clinicians there saw 52 patients who had neurosarcoidosis instead of MS [13]. The diagnosis of sarcoidosis is generally made by demonstrating inflammation on biopsy of affected lymph nodes, liver, nerve, or skin, with characteristic non-caseating granulomas. MRI scans of the brain usually reveal meningeal enhancement which is

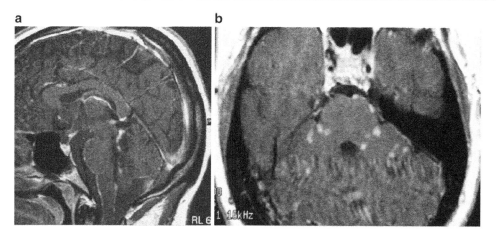

Fig. 6.2 MRI of the brain in neurosarcoidosis. (**a**) Sagittal and (**b**) axial post-gadolinium images demonstrating thick, irregular, and nodular leptomeningeal enhancement, which is more striking along the basilar and perimesencephalic cisterns, affecting the cisternal segments of the cranial nerves. Note the enhancing nodules along the interpeduncular, crural, and pre-pontine cisterns along the course of cranial nerves III, VI, V, and VI, respectively

frequently nodular, or diffuse intraparenchymal inflammatory lesions (Fig. 6.2). Alternatively, there may be diffuse white matter lesions indistinguishable from MS as in the patient in Inset 6.5.

Inset 6.5 Neurosarcoidosis Mimicking MS—A 49-Year-Old Woman with White Matter Lesions and a History of Hearing Loss

EF was a 49-year-old woman referred to our MS center because of an MRI scan read as likely multiple sclerosis.

She was well until 19 years prior to the evaluation. Shortly after delivering her first child she developed progressive complete loss of hearing on the left side over a few weeks and underwent an MRI scan of the brain. This showed multiple white matter lesions, and she was told it was possible MS. She did not return to the neurologist for follow-up. The hearing loss slowly improved without treatment, but she remained with mildly decreased hearing on the left.

Ten years prior to the evaluation she began to experience episodes of severe arthritis lasting a few weeks to a few months, consisting of knee or ankle swelling, pain, and limitation of activity. These episodes were treated with nonsteroidal anti-inflammatory drugs (NSAIDs) with some relief, and ultimately resolved without recurrence.

About 5 years prior to this evaluation, she developed pain and difficulty with vision in both eyes, diagnosed as uveitis, and was treated with eye injections, eye drops, and oral prednisone. Over the year prior to this evaluation, she has had worse headaches than usual, responding partially to NSAIDs. Because of the headaches, she underwent an MRI of the brain which revealed multiple white matter lesions, read as highly suggestive of MS. None of the lesions enhanced with gadolinium injection. She had not noted any significant numbness or weakness, gait unsteadiness, or any other neurological symptoms other than the difficulty with vision which had been attributed to uveitis. Neurological evaluation was negative except decreased vision.

Initial evaluation was felt to be consistent with sarcoidosis, and she underwent a gallium scan which revealed inflammation in her minor salivary glands and lacrimal glands. These were biopsied, revealing inflammation

(continued)

Inset 6.5 (continued)

and noncaseating granulomas, confirming the diagnosis. Chest X-ray showed hilar adenopathy. A lumbar puncture revealed a lymphocytic pleocytosis and no oligoclonal bands. For her uveitis and headache, she was treated with methotrexate and alternate day corticosteroids.

Over a follow-up of 4 years, her uveitis and meningitis improved and she was able to be taken off the medications. Her MRI of the brain was repeated 4 years after initial evaluation and showed no increase in white matter lesions.

Author's note. Most MS patients are quite healthy on presentation, with the exception of their problems in the central nervous system. This patient had had inflammation in her joints and in her uvea, problems not usually associated with MS, although an occasional MS patient can have an associated uveitis. In addition, she had hearing loss as an initial problem, a symptom which is very unusual in MS. The biopsy confirmed that the patient had an alternative explanation for her symptoms.

6.2   Subacute Combined Degeneration

This is a rare but highly treatable cause of demyelination of the dorsal and lateral columns of the spinal cord, caused by vitamin B12 deficiency. This is usually due to pernicious anemia, an autoimmune disease in which there are antibodies to the parietal cells of the stomach, leading to inadequate secretion of intrinsic factor, a glycoprotein required for vitamin B12 absorption in the intestines. As the vitamin B12 levels drop, patients usually develop sensory deficits, weakness, ataxia and postural instability, and sphincter disturbances. More acute presentations of days or a week or two have been described. The pathology is usually in the posterior columns and lateral corticospinal

tracts [14]. The disease usually does not fluctuate, and although rare, it always must be considered in the differential of MS because it is completely treatable by supplementation with vitamin B12.

6.3   Other Neuroinflammatory Diseases

Some diseases are so rare that they are normally only considered far down the list of likely suspects. These are sometimes called "zebras"; a zebra is an extraordinarily rare disease presentation that has an extremely low probability of being the ultimate diagnosis in a certain presentation and comes from the phrase: "When you hear hoofbeats, think horses, not zebras." Some neuroimmunological rare diseases that have importance beyond their prevalence will be discussed in Chap. 16. "Zebras" that have been very rarely confused with MS include inflammatory diseases (Sjogren's syndrome, Behcet's disease, isolated CNS vasculitis and systemic vasculitides such as Wegener's granulomatosis or Churg–Strauss vasculitis) and diseases without inflammation (CADASIL and Susac's syndrome). Although these diseases may have characteristics that resemble MS in an acute presentation, their natural history is generally quite different from MS. The neuroinflammatory diseases will be described in other chapters, and CADASIL and Susac's will be discussed below.

CADASIL is the acronym for cerebral autosomal dominant arteriopathy with subcortical infarcts and leukoencephalopathy. It is the most common heritable cause of stroke and vascular dementia in adults and is associated with a mutation in the Notch 3 gene. This gene encodes for a transmembrane receptor, and the mutations causing CADASIL result in an odd number of cysteine repeats in that protein. It is felt that the underlying pathology is a chronic subcortical ischemia: Notch 3 forms aggregates around cerebral blood vessels. The patient referred to our MS center described in Inset 6.6 was not unusual in that the disease is frequently associated with migraine with auras and with accumulating subcortical white matter lesion burden.

Inset 6.6 Multifocal White Matter Disease on MRI Scanning in a Patient with Frequent "Migraines"

MH was a 37-year-old right-handed woman when she was referred to our MS center by her local neurologist for the question of the diagnosis of MS.

She had sought his attention a few years prior when she began having increasingly frequent episodes of what she called her "migraine," a severe throbbing headache preceded by a visual image of gold spinning in her eye, and numbness and tingling in her arms and legs. These occurred about once a month. The sensations of numbness and tingling never occurred without the headaches. She had several MRIs of the brain which initially showed a small number of white matter lesions, most of them subcortical and not periventricular, but then an increasing number over time; none ever enhanced with gadolinium. The rest of the history was negative, including the absence of stroke risk factors, except that the family history was positive for strokes and migraines in a number of relatives. Her neurological examination was normal. Evoked responses were negative for lesions consistent with MS and the spinal fluid examination was normal with negative oligoclonal bands. Because of the strongly positive family history, the accruing white matter lesions, and the frequent migraines with aura, CADASIL was considered. An assay for a mutation in the Notch 3 gene on chromosome 19 was positive. She was referred to a CADASIL specialist in New York City for further workup, genetic counseling, and therapy.

Author's note. Usually migraine headaches begin during teenage years or even younger, and the onset of migraine episodes in one's 30s is unusual. Thus, idiopathic migraine would have been possible but not likely. The accrual of new lesions over time, their subcortical rather than periventricular location and the absence of oligoclonal bands on the spinal tap all supported the diagnosis of CADASIL, despite the fact that CADASIL is much less common than MS. CADASIL is rare, and this was the only patient I have ever cared for in whom I have made that diagnosis.

Susac's syndrome. Susac's syndrome (SS) consists of the clinical triad of encephalopathy, branch retinal artery occlusions, and hearing loss [15]. The disease is caused by a small blood vessel disease affecting the eyes, but the underlying etiology is unknown, and it is not clearly neuroinflammatory. It may be post-viral since many cases spontaneously improve, but the recommended therapy is immunosuppression for the acute inflammatory syndromes.

References

1. Costello DJ, Eichler AF, Eichler FS. Leukodystrophies: classification, diagnosis, and treatment. Neurologist. 2009;15(6):319–28.
2. Tumani H, Sapunova-Mayer I, Sussmuth SD, Hirt V, Brettschneider J. CIS case studies. J Neurol Sci. 2009;287 Suppl 1:S7–10.
3. Brain WR, Walton JN. Diseases of the nervous system. 7th ed. New York: Oxford University Press; 1969.
4. Murray TJ. Multiple sclerosis: the history of a disease. 1st ed. New York: Demos; 2005.
5. Pachner AR, Steiner I. Lyme neuroborreliosis: infection, immunity, and inflammation. Lancet Neurol. 2007;6(6):544–52.
6. Pandit L. Differential diagnosis of white matter diseases in the tropics: an overview. Ann Indian Acad Neurol. 2009;12(1):12–21.
7. Collongues N, Marignier R, Zephir H, et al. Neuromyelitis optica in France: a multicenter study of 125 patients. Neurology. 2010;74(9):736–42.
8. Bennett JL, Lam C, Kalluri SR, et al. Intrathecal pathogenic anti-aquaporin-4 antibodies in early neuromyelitis optica. Ann Neurol. 2009;66(5):617–29.
9. Bradl M, Misu T, Takahashi T, et al. Neuromyelitis optica: pathogenicity of patient immunoglobulin in vivo. Ann Neurol. 2009;66(5):630–43.
10. Jacob A, Matiello M, Weinshenker BG, et al. Treatment of neuromyelitis optica with mycophenolate mofetil: retrospective analysis of 24 patients. Arch Neurol. 2009;66(9):1128–33.

11. Wingerchuk DM, Weinshenker BG. Neuromyelitis optica. Curr Treat Options Neurol. 2008;10(1): 55–66.

12. Dale RC, de Sousa C, Chong WK, Cox TC, Harding B, Neville BG. Acute disseminated encephalomyelitis, multiphasic disseminated encephalomyelitis and multiple sclerosis in children. Brain. 2000;123(Pt 12):2407–22.

13. Pawate S, Moses H, Sriram S. Presentations and outcomes of neurosarcoidosis: a study of 54 cases. QJM. 2009;102(7):449–60.

14. Vasconcelos OM, Poehm EH, McCarter RJ, Campbell WW, Quezado ZMN. Potential outcome factors in subacute combined degeneration. J Gen Intern Med. 2006;21:1063.

15. Susac JO, Egan RA, Rennebohm RM, Lubow M. Susac's syndrome: 1975-2005 microangiopathy/autoimmune endotheliopathy. J Neurol Sci. 2007;257 (1–2):270–2.

16. Vickrey BG, Samuels MA, Ropper AH. How neurologists think: a cognitive psychology perspective on missed diagnoses. Ann Neurol. 2010;67(4):425–33.

Multiple Sclerosis Therapy

7

Lacking proof of the value of any form of therapy in multiple sclerosis, the average physician or neurologist is usually content to rely on a placebo.

Douglas McAlpine, Multiple Sclerosis—A Reappraisal, 1965, p. 197

In long-term diseases, it can be difficult to execute randomized clinical trials (RCTs) which reach hard unambiguous efficacy endpoints.

Ebers et al. (2008) [1]

The history of multiple sclerosis therapy is generally one of futility, consisting of many far-fetched remedies, some of which have been painful or harmful or both. This is not surprising given the variability of the disease among patients, and even within the same patients, and considering the relative inability of current therapy to reverse the disease. However, significant advances have been made in MS therapy, which have made McAlpine's pessimistic statement in 1965 outdated. The pessimism of previous times has been replaced by a flurry of attempts to optimize modern clinical trial design to identify effective treatments. These attempts, which have come from a massive amount of investment by pharmaceutical companies and the extensive efforts of neurologists and allied health care workers, have led to the discovery of many effective therapies. These therapies can be divided into two general categories. This chapter first covers "disease-modifying drugs (DMDs)," therapies which affect the underlying disease process. Next "symptomatic therapies" are discussed, including life-style changes which can benefit the patient.

1 Disease-Modifying Drugs

1.1 How Medications Become Approved for Use in MS

1. *Preclinical studies.* Most large pharmaceutical companies will have a department devoted to drug discovery, in which research is performed to identify possible therapies. These studies are frequently performed in in vitro systems such as cell cultures, and will often include studies in animal models. Animal models for MS will be extensively discussed in Chap. 8.
2. *Phase 0 studies.* These are early pilot studies of a new drug in a small number of human volunteers. They address relatively simple issues of pharmacokinetics (PK) and pharmacodynamics (PD), usually after only one administration of the drug.
3. *Phase 1 studies.* These are usually called safety and dose-finding studies, although they also are designed to obtain more PK/PD data than could be obtained in Phase 0. Usually, low doses are initially used followed by

A.R. Pachner, *A Primer of Neuroimmunological Disease*,
DOI 10.1007/978-1-4614-2188-7_7, © Springer Science+Business Media, LLC 2012

increasing amounts of drugs as low doses are found to be safe.

4. *Phase 2 studies.* Whereas phase 0 and 1 involve small numbers of subjects, phase 2 generally involve larger numbers for longer periods to try to address three issues: dosing, safety, and efficacy.

5. *Phase 3 studies.* These are large, very expensive trials which establish the drug's efficacy and safety compared to either placebo or a gold standard treatment. They are randomized, controlled trials (RCTs). RCTs have been a major step forward for MS and have addressed the problems with bias, blinding, and statistical analysis of earlier "positive" studies of MS therapies. Given the inherent variability of MS, difficulties in measurement, and strong placebo effects, RCTs are critical to assess efficacy of a new agent; they have now become the gold standard. These studies generally incorporate the following characteristics:

   (a) *They are randomized.* When patients enter the study, they are randomly assigned to one of the experimental arms, rather than choosing one of the arms or being assigned by a study physician.

   (b) *They are double-blinded.* Neither the patient nor the study physicians know which arm of the study the patient has been randomized to. Occasionally, the term triple-blind is used for some studies in which an additional layer of individuals in the study is masked as to the treatment groups. For instance, some clinical trials specialists recommend that statisticians analyzing data be blinded, since potential bias can be present in the analysis of the study.

   (c) *They are placebo-controlled.* At least one of the arms involves treatment with a placebo, an agent that is not biologically active and has no effect on the disease process. More recently, with the advent of a number of partially effective agents, more RCTs are beginning to be active comparator-controlled, rather than placebo-controlled. An active comparator is an agent that is currently being used for MS that is known to be effective. One reason for employing an active comparator is that it is easier to recruit MS patients into a study when they know they will not receive a placebo. Another reason is the hope that the new drug will be superior to the old drug, and thus capture a large market share. An example is the recently published TRANSFORMS study in which patients were randomized to fingolimod or interferon-beta-1a (IFN-β-1a), and a placebo arm was not included [2].

A positive aspect of the change to using RCTs is that success of a new agent in an RCT gives considerable confidence that the drug is indeed effective. A negative effect is that RCTs are very expensive. A result of this expense of the clinical trials is that new drugs are tested for 2 years or less. This very short period of testing can be adequate for testing for decreasing relapse rates but precludes the ability to test for effects on disability progression [1]. Another consequence of the great expense of RCTs is that new drugs tend to be very expensive. Despite the above negatives, the MS community sees RCTs as a major advance in developing new therapies for MS.

1.2   The Primary Endpoint

The disadvantages of RCTs, primarily their expense and length, are outweighed by their scientific rigor in establishing a reliable answer to the question: does this drug work in MS? But, as is the usual in clinical issues, the devil is in the details. What does the word "work" mean; in other words, what aspect of MS responds in a positive way to the therapy? All RCTs must have a primary endpoint, a single measure of the disease in which the MS population being treated with the drug is compared to either a placebo-treated population or an active comparator. This must be a measure which can be reduced to a number. This requirement for a primary endpoint is due to the need, prior to study recruitment, to identify the number and kind of patients required, and the needed time in the study.

Pharmaceutical companies formulate the primary endpoint according to what will allow their product to be approved by regulatory agencies such as the FDA in the USA and EMEA in Europe, not according to what neurologists, patients, patient families, radiologists, or pathologists consider to be most important in the disease. As discussed briefly in Chap. 4, identifying a single aspect of MS to focus on is fraught with conceptual peril. Up to now, the disease-modifying agents have been approved generally on the basis of trials demonstrating a reduction in relapse rate. Many neurologists feel that future drugs should demonstrate efficacy in reducing disability progression to achieve approval, since our armamentarium for treating relapse frequency is now reasonably adequate. However, there is no consensus about how to measure disability as a primary endpoint in a drug trial. This is because of the slow accrual of disability over time in most MS patients and the inherent problems with disability measurements in short-term studies.

1.3 When Should One Treat with DMDs?

The pivotal studies of the DMDs in the 1990s were performed in patients who had clinically definite MS, often for at least a year. When these agents were shown to be effective in this population, subsequent studies were performed in CIS patients to see if DMDs could ameliorate the course of the disease when administered earlier in the disease. The first such study was the CHAMPS study, in which Jacobs et al. demonstrated that, in patients with CIS and at least two brain MRI lesions typical of MS (a population highly likely to develop clinically definite MS) 35% of those treated with intramuscular IFN-β-1a developed CDMS, while in the placebo group, 50% developed CDMS [3]. This difference was highly significant and led to the widespread use of DMDs in patients with CIS and brain MRI lesions typical of MS. Subsequent studies with IFN-β-1b and GA have confirmed these results. Thus, most neurologists now treat

with DMDs as soon as possible in patients with MS or with a clinical presentation highly likely to develop into MS.

1.4 General Approach to Therapy

1. The goal of current therapies is the reduction of relapses or gadolinium-enhanced lesions on MRI, which is hoped to result in delayed or reduced disability.

As mentioned above, the primary endpoint for most studies demonstrating efficacy of MS drugs has been reduction in relapses. One of the "platform drugs," as discussed below, is thus indicated for patients with relapses, especially when the relapses are frequent. However, it is not a given that a medication that will lower relapse frequency will also definitively slow disease progression. In a study of 1,844 MS patients, Confavreux et al. came to the following conclusions.

> We found that once a clinical threshold of irreversible disability has been reached (a score of 4 on the Kurtzke Disability Status Scale), the progression of disability is not affected by relapses, either those that occur before the onset of the progressive phase or those that supervene during this phase. The absence of a relation between relapses and irreversible disability suggests that there is a dissociation at the biologic level between recurrent acute focal inflammation and progressive degeneration of the central nervous system. This apparent paradox is consistent with the persistence of the progression of disability in patients with multiple sclerosis despite infection with the human immunodeficiency virus or despite suppression of the cerebral inflammation after treatment with a potent antileukocyte monoclonal antibody. It also suggests that agents that have a short-term effect on relapses in patients with multiple sclerosis may not necessarily delay the development of disability in the long term. [4]

Their findings were confirmed in a more recent large study by Scalfari et al. [5] utilizing the Sylvia Lawry database. Their conclusion was that neither total number of attacks nor attacks experienced after the second year of the disease correlated with disability progression. It is thus our *hope* rather than a proven *fact* that available therapies for MS, especially when started early in the disease course, are effective in delaying or suppressing disability progression.

1.5   Current Agents

1.5.1   Corticosteroids

The most commonly used corticosteroid for acute attacks of MS is methylprednisolone, also called Solumedrol. This compound has a molecular weight of 374 Da and is a typical corticosteroid with a structure derived from cholesterol with multiple sites of hydroxylation. Corticosteroids are used by neurologists to treat an ever increasing list of diseases. They are generally safe over the short term. However, large doses, especially given over an extended time period, can cause many adverse effects, including elevated blood sugars and diabetes, decreased bone density (osteoporosis), psychiatric complications (including anxiety, depression, and sleeping disorders), hypertension, and cataracts. For this reason, the chronic use of corticosteroids at high doses is not considered an acceptable therapy for MS. Instead, corticosteroids are generally used to treat attacks, at which time they are administered at high dose intravenously or orally for a short time period, usually just 3–5 days.

One of the first randomized, clinical trials related to MS was testing of corticosteroid therapy for optic neuritis, called the Optic Neuritis Treatment Trial (ONTT) [6]. The entry criteria required relatively severe optic neuritis with both an afferent pupillary defect (APD) and a visual field defect, but did not require the presence of MS, and excluded patients with already known MS. The study found that intravenous treatment with one gram of methylprednisolone (Solu-Medrol), per day for 3 days accelerated visual recovery but did not improve visual outcome after 1 year. The study also found that just the one treatment with IV Solumedrol therapy delayed the development of MS relative to untreated individuals. This multicenter study laid the groundwork for many of the large RCTs of other agents for MS in the subsequent decades.

1.5.2   First-Line Disease-Modifying Drugs in MS

Interferon-β (Brand Names: Avonex, Betaseron, Rebif, Extavia)

> How does IFN-β work in MS? We haven't got a clue.
>
> Mark Freedman, University of Ottawa, Ottawa, Canada, 2010

When MRI became increasingly used for MS diagnosis, it became clear that MS was not just a disease of clinical attacks with normalcy in between attacks, but rather a constantly present disease in which lesions were continually accrued, most of which occurred subclinically. IFN-β is a treatment which is used continually even in the absence of attacks.

An extended history of the development of IFN-β is beyond the scope of this primer. The drug was initially used intrathecally on the theory that MS was a viral infection and might respond to IFN-β, which has strong antiviral effects. Later, the route of administration was switched to systemic injection, i.e., subcutaneous or intramuscular. IFN-β is in a class of proteins called interferons, of which there are many different forms in human. The most commonly studied interferons are labeled IFN-α, -β, and -γ. IFN-β is a glycoprotein of approximately 20,000–22,000 Da in size and is considered a biological therapy; like most biological therapies, is produced by recombinant technology. There are currently four preparations of IFN-β available in the USA and Europe which can be divided into IFN-β-1a or IFN-β-1b; these differences are briefly summarized in the accompanying table (Fig. 7.1). The IFN-β-1a products are made in genetically engineered Chinese hamster ovary (CHO) cells, have the same amino acid structure as does human IFN-1β, and are glycosylated, while the IFN-β-1b products are made in the bacterium *E. coli*, have one amino acid difference from natural IFN-β-1, and are not glycosylated.

The mechanism of action of all of the IFN-β preparations is binding to the interferon α–β receptor (IFNAR) leading to the upregulation or downregulation of a hundreds of genes, as shown in Fig. 7.2. Many cells in the body have IFNARs and are thus susceptible to the IFN-β effects, but the specific target cell types and the up- or downregulated genes responsible for the drug's effect in MS are unknown. Many of the genes regulated by IFN-β have roles in innate immunity and protect cells against destruction during a viral attack. An example of such a gene is the Mx1 gene, the protein product of which has been used extensively for detecting IFN-β bioactivity in IFN-β-treated MS patients [7]. IFNAR activation is

	Betaseron® (interferon beta-1b)	Avonex® (interferon beta-1a)	Rebif® (interferon beta-1a)*	
Molecular weight	18.5 Kd (165 amino acids)	22.5 Kd (166 amino acids)	22.5 Kd (166 amino acids)	
Class of molecule	Protein	Glycosylated protein	Glycosylated protein	
Derived	Human fibroblasts	Human fibroblasts	Human fibroblasts	
Host Cell	*E coli*	Chinese Hamster Ovary	Chinese Hamster Ovary	
Amino acid sequence	Serine substitute for Cysteine at position 17	Identical to Human	Identical to Human	
Administration	Subcutaneous	Intramuscular	Subcutaneous	
Approved Dose	250 mcg (0.25 mg) every other day	30 mcg (0.03 mg) once a week	44 mcg (0.044 mg) three times a week*	22 mcg (0.022 mg) three times a week*
Specific Biological Activity	32 MIU/mg World Health Organization Standard	200 MIU/mg World Health Organization Standard	200 MIU/mg World Health Organization Standard	200 MIU/mg World Health Organization Standard
Biological Activity of Approved Weekly Dose	250 mcg x 32 MIU/mg 28 MIU†	30 mcg x 200 MIU/mg 6 MIU†	44 mcg x 200 MIU/mg* 26.4 MIU†	22 mcg x 200 MIU/mg* 13.2 MIU†
Total Weekly Dose	250 mcg x 3.5 doses/week = 875 mcg weekly .875 mg x 32 MIU = 28 MIU 3.5 doses/week x 8 MIU = 28 MIU or 875 mcg contains 28 MIU of biological activity	30 mcg x 1 dose/week = 30 mcg weekly .030 mg x 200 MIU = 6 MIU 1 dose/week x 6 MIU = 6 MIU or 30 mcg contains 6 MIU of biological activity	44 mcg x 3 doses/week = 132 mcg weekly .044 mg x 200 MIU = 8.8 MIU 3 doses/week x 8.8 MIU = 26.4 MIU or 132 mcg contains 26.4 MIU of biological activity	22 mcg x 3 doses/week = 66 mcg weekly .022 mg x 200 MIU = 4.4 MIU 3 doses/week x 4.4 MIU = 13.2 MIU or 66 mcg contains 13.2 MIU of biological activity

Fig. 7.1 Types of interferon-β currently available for therapy of MS. Recently, Extavia, an IFN-β-1b product has been made available. This drug is identical to Betaseron, the drug in the first column of the table

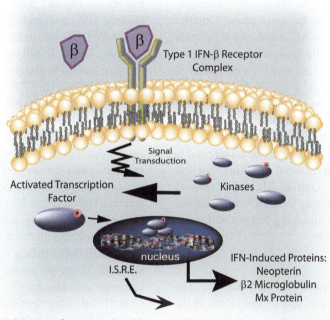

I.S.R.E. = Interferon Stimulated Response Element

Fig. 7.2 IFN-β signal transduction. IFN-β, shown as "β" in this cartoon, works by binding to a specific receptor on the cell surface, shown as the "Type 1 IFN receptor complex," but also known as the interferon-α/β receptor or IFNAR. This receptor binds either IFN-β or IFN-α. After binding of IFN-β to the receptor, there is an orderly sequence of intracellular events including activation of kinases and transcription factors, and after hours upregulation of mRNAs for a host of IFN-upregulated genes required for IFN-β effect, and MxA and other IFN-β-inducible genes can serve as biomarkers of this effect.

Therapeutic Effect

Multiple studies have demonstrated that IFN-β has at least two salutary effects in MS: a decrease in attack frequency and a decrease in the occurrence of gadolinium-enhancing lesions. These effects are most obvious in the first year of therapy but less obvious after that. These drugs along with glatiramer acetate (see below) have become the standard of care for MS therapy. However, some investigators, using careful analysis of the multiple clinical trials, have called into question the magnitude of the effect: "Recombinant interferons slightly reduce the number of patients who have exacerbations during first year of treatment. Their clinical effect beyond 1 year is uncertain and new trials are needed to assess their long-term effectiveness and side-effects [8]."

The effect of IFN-β on disability progression over an extended period of time has not been carefully studied. In most MS patients, disability accrues slowly, so that the usual phase 3 studies lasting 2 years or less will not detect significant differences in disability between treated and placebo groups, even if an ameliorative effect is present. One of the problems with these short studies is that patients accrue temporary disability from attacks, which sometimes lasts many months. Consequently, in a short study, a salutary effect on relapse frequency may be manifested in an effect on relapse-induced disability, but this effect may be quite transient. Most clinicians are hopeful that over the long term decreased relapses and decreased gad-enhancing lesions will translate into less disability.

Side Effects

IFN-β is well tolerated by most patients. The most common problem, usually felt mostly in the first few months of therapy, is "flu-like symptoms." Within a few hours after the injection, many patients experience malaise, muscle and joint aches, low-grade fever, and increased fatigue. Other patients never develop these symptoms. A very low percentage of patients develop low-grade liver injury, which is rapidly reversible upon discontinuation of the medication. The drug should not be injected during pregnancy. Skin reactions occur relatively frequently with subcutaneous preparations, but these are usually not severe enough to require discontinuation of drug. Mild decreases of the concentration of white blood cells is the peripheral blood are common, but almost never a cause for stopping the drug.

Other Issues Affecting Use of IFN-β in MS

– *Anti-IFN-β antibodies.* IFNAR activation, which is the pathway by which the drug, is compromised in some patients by anti-IFN-β antibodies, which bind to circulating IFN-β after injection and prevent its binding to IFNAR. Because IFN-β is injected at frequent intervals, ranging from every other day to weekly, the injections serve as a nearly constant antigenic challenge, and antibodies, when they develop, usually begin to increase during the first months of therapy. These antibodies can block the effects [9] and diminish the therapeutic efficacy [10]. Although therapeutic IFN-β is manufactured by pharmaceutical companies to resemble human IFN-β as closely as possible to minimize its immunogenicity, all of the current preparations have differences from human natural IFN-β which the human immune system can discern. Two different assays can be used to measure anti-IFN-β antibodies in IFN-β-treated patients. The first is a fast, inexpensive enzyme-linked immunosorbent assay (ELISA) which measures all antibodies in the human serum which bind to IFN-β. If this assay is positive, a neutralizing antibody (NAb) assay is performed, in which the ability of antibody to block IFN-β binding to IFNAR is tested in vitro. These antibodies do not occur in untreated individuals.

– *Cost.* IFN-β is an expensive agent, with costs in the USA usually being between 20,000 and 30,000 dollars per patient per year. This has led to a call for generic IFN-β products which would be much less expensive. However, quality control is a major concern for generic biologics, in general, and generic IFN-β preparations would need to demonstrate comparable high potency and low immunogenicity.

- *Adherence to therapy.* The fact that IFN-β needs to be administered by injection between every other day to once a week, with usually no change in the disease perceptible to the patient, has resulted in less than optimal compliance with these drugs. The pharmaceutical companies making these products have attempted with variable success to address this issue, but studies have demonstrated that many and possibly most patients are not fully adherent to the injection protocols. It is theoretically possible to assess this by measuring IFN-β-upregulated biomarkers in blood, such as the MxA protein [11], but these assays are not routinely available for clinical use.

Glatiramer Acetate (GA)

GA was initially developed at the Weizmann Institute in Israel to be used as an aid to researchers in the induction of experimental autoimmune encephalomyelitis (EAE), an experimental model of post-vaccinial encephalomyelitis sometimes used as a model of MS (see Chap. 8). Although it was unable to induce EAE, it was found to interfere with the induction of EAE [12]. The basis for the biological activity of GA was felt to be its immunological cross reactivity with myelin basic protein (MBP), the "encephalitogen" in those experiments. GA is a synthetic random copolymer of L-alanine, L-glutamic acid, L-lysine, and L-tyrosine in a residue molar ratio of 6.0:1.9:4.7:1.0.

For the treatment of MS, 20 mg of GA is injected subcutaneously daily. The mechanism of action of GA in MS is unknown and although it was developed as a congener for MBP to suppress anti-MBP-mediated autoimmunity, there is no evidence for anti-MBP autoimmunity in MS. Many mechanisms have been postulated, but none have been able to be broadly reproduced or utilized for monitoring therapy in practice. Some of its effects may be due to the large protein challenge injected subcutaneously every day, which may generate a constant immunological challenge; this hypothesis is supported by persistent anti-GA antibodies [13] in GA-treated patients. Unlike IFN-β, which operates by binding to a well-characterized membrane receptor thereby regulating a large number of genes, and thus can be followed by a number of biomarkers, GA has no receptor or consistently upregulated genes or biomarkers.

Therapeutic Effect

MS patients who injected glatiramer in pivotal phase 3 studies experienced fewer relapses and developed fewer gadolinium-enhanced lesions on brain MRI than those who injected placebo. As in therapy with IFN-β, an effect of therapy on disability progression has not been proven.

Side Effects

GA is usually well-tolerated. Rarely, patients can get a reaction within ½ h after injection consisting of flushing, shortness of breath, palpitations, and chest tightness. Other patients can develop pitting of skin at sites of injection due to the loss of fat called lipoatrophy. The incidence of lipoatrophy is unclear from the literature, ranging from rare to nearly 50% of patients.

Other Issues

- *Anti-GA antibodies.* Daily subcutaneous injection with GA induces anti-GA antibodies in nearly all patients. However, because there is no receptor or biomarker for GA, neutralizing antibody assays cannot be performed. The requirement for an NAb assay is that the treatment must induce a reproducible, measurable in vitro effect which could then be abolished by NAbs, and such an in vitro effect has not yet been found for GA. Unlike the situation in IFN-β therapy, in which anti-IFN-β antibodies interfere with the therapeutic effect, anti-GA antibodies are felt to have no adverse effects and were actually found at higher levels in relapse-free patients [14].
- *Cost.* GA's expense, like that of IFN-β, is high.
- *Adherence to therapy.* As with IFN-β, adherence to therapy is not optimal. With GA treatment requiring daily injections, most GA-treated patients are not fully adherent to therapy as

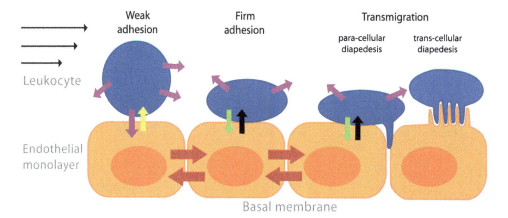

Fig. 7.3 Migration of a lymphocyte out of the blood-stream into a lymph node. A circulating lymphocyte adheres weakly to the surface of the specialized endothelial cells lining a postcapillary venule in a lymph node. This initial adhesion is mediated by L-selectin on the lymphocyte surface. The adhesion is sufficiently weak to enable the lymphocyte to roll along the surface of the endothelial cells, pushed along by the flow of blood. Stimulated by chemokines secreted by the endothelial cells, the lymphocyte rapidly activates a stronger adhesion system, mediated by an integrin. This strong adhesion enables the cell to stop rolling and migrate out of the venule between the endothelial cells. The subsequent migration of the lymphocytes in the lymph node also depends on chemokines, which are produced within the node. The migration of other white blood cells out of the bloodstream into sites of infection occurs in a similar way

prescribed. It is unclear how strictly adherent patients must be to receive benefit from the drug.

1.5.3   Second-Line Drugs

These drugs are usually used in the USA when the platform drugs have proven ineffective. The particular conditions that define "ineffective" will be highly variable from patient to patient and from neurologist to neurologist. These drugs are considered second level, not first line, because there are known serious adverse events (SAEs) (e.g., natalizumab or mitoxantrone) or because the drugs have not been clearly demonstrated to be effective in a RCT (e.g., immunosuppressives such as methotrexate or cyclophosphamide). In Europe, many drugs which are felt to be second line in the USA, such as azathioprine, have been popular as first-line agents [15]. A head-to-head study of azathioprine vs. IFN-β has been recently proposed [16]. The lack of a clearly optimal treatment protocol is a testament to the lack of full efficacy of any one agent or combination of agents.

Natalizumab (Brand Name: Tysabri)

Natalizumab is the first therapeutic monoclonal antibody demonstrated to be effective in MS. Natalizumab is a humanized neutralizing IgG4 kappa monoclonal antibody against α4 integrins (see Sect. 2.2.5 of Chap. 1 for a brief introduction to adhesion molecules). α4β1 integrin, a heterodimer (composed of an alpha and a beta chain), is also occasionally referred to as "very late antigen-4 (VLA-4)," since it is expressed relatively late in the inflammatory response. Cells bearing α4β1 integrin can bind to another cell adhesion molecule called VCAM-1, also known as CD106, which is expressed on endothelial cell membranes only after exposure to cytokines. As an anti-α4 integrin antibody, natalizumab also targets cells which have α4β7, a molecule essential for mononuclear cell influx into the small intestine. Integrins are felt to be critical to allow white cells to enter tissues from the circulation. This involves a progressive slowing of the velocity of the cell, followed by rolling, and then adhesion and migration, as shown in Fig. 7.3. The ability of natalizumab to interfere with leukocyte entry into the CNS by binding to and blocking α4β1 integrin on the surface of leukocytes is what is thought to mediate its salutary effects in MS [17]. However, α4 integrin is a molecule present throughout the immune system, and other mechanisms, either in addition to or instead of CNS entry, are equally plausible. Thus, the precise mechanisms for its effects in MS are not known. Because of the relatively long half life of IgG4, natalizumab is infused once per month.

The current product insert for the drug, which is an FDA-regulated summary of usage, states that "Tysabri is generally recommended for patients that have not been helped enough by, or cannot tolerate, another treatment for MS." That is, Tysabri is recommended as a second-line agent after the use of one of the platform drugs. The reason is the occurrence of a SAE. After initially being available in November, 2004, natalizumab was pulled from the marketplace in February, 2005, only 3 months later, because three participants in clinical trials of the drug developed progressive multifocal leukoencephalopathy (PML), a dangerous and frequently fatal viral CNS infection. Since then, natalizumab has been reintroduced into the marketplace with a number of precautions. The judgment of the FDA was that the drug was very effective in MS and could be used as a second-line agent. This hypothesis is currently being tested in the SURPASS study, which will test the performance of Tysabri against IFN-β or glatiramer. The development of PML is a rare side effect of a number of immunosuppressive medications, as well as being a not uncommon manifestation of severe immunosuppression in HIV-AIDS. PML will be discussed at greater length in Chap. 12. A serum anti-JC virus antibody assay to test for previous exposure to the JC virus has recently become available to clinicians to assess the risk of PML in a patient. Patients who are seronegative are at less risk for PML than patients who are seropositive; studies on this anti-JC virus antibody assay are ongoing.

Another problem in natalizumab therapy of MS is neutralizing antibodies to the therapy which result in the loss of efficacy. This is less of a problem for this therapy, in contrast to IFN-β, and is found in only 6% of recipients of the drug.

Mitoxantrone (Novantrone)

Mitoxantrone, which was previously used primarily as an anticancer drug, was approved by the FDA in the USA in 2000 for the treatment of worsening relapsing–remitting, secondary progressive, and progressive relapsing MS, based on the Mitoxantrone in MS (MIMS) study [18]. Like natalizumb, it is also considered to be an agent not to be used as a platform drug in patients with newly diagnosed relapsing–remitting MS. It has a demonstrated effect in an underserved population of MS patients, those with more severe disease or who are progressing in disability [19].

Mitoxantrone is an anthracenedione compound that interferes with DNA synthesis and is primarily used as an antineoplastic agent. It also is immunosuppressive, affecting almost all mononuclear cells including T cells, B cells, and macrophages. Adverse effects initially were thought to be primarily related to its cardiotoxicity which could be avoided by careful noninvasive cardiac monitoring by echocardiography. However, recently increasing reports of treatment-related acute leukemias (TRAL) in MS patients treated with mitoxantrone have led to more concern about the risk–benefit ratio for mitoxantrone. A recent report by a Therapeutics and Technology Assessment Subcommittee of the American Academy of Neurology concluded that "clinicians contemplating MX administration for an individual patient with MS must weigh the potential for benefit against the potential for harm given the (approximately) 12% risk of systolic dysfunction and (approximately) 0.8% risk of TRAL and the availability of alternative therapies with less severe toxicities (e.g., IFN-β and glatiramer acetate) for patients with RRMS" [20].

1.5.4 Old Drugs for MS
Immunosuppressives

Azathioprine (Imuran), cyclophosphamide (Cytoxan), and methotrexate (Rheumatrex) are drugs used as immunosuppressives in MS and are discussed in Chap. 19.

Controversies

The use of immunosuppressives in MS is controversial. Azathioprine is considered a relatively

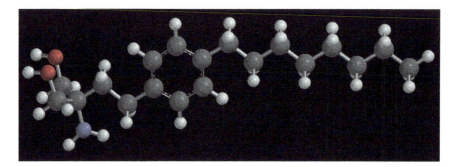

Fig. 7.4 Drawing of chemical structure of fingolimod (Wikimedia, public domain at http://commons.wikimedia.org/wiki/File:Fingolimod.png)

safe immunosuppressive drug in Europe and is commonly used there. A recent large meta-analysis of the use of azathioprine in MS comes to a favorable conclusion and recommends a comparison study of IFN-β with azathioprine [16]. In some MS practices, the more potent cyclophosphamide at moderate doses is used extensively as a platform drug, while in most MS centers, it is used rarely, and only in patients with highly inflammatory presentations. Two trials published in the early 1990s came to different conclusions: results from the Northeast Cooperative Multiple Sclerosis Treatment Group was supportive of its use, while a Canadian study found no significant between-group differences in time to treatment failure. Various aspects of the controversies over the use of immunosuppressive agents in MS are summarized in a recent review article [21].

1.5.5   New Drugs for MS

Two oral medications have demonstrated significant benefit in reducing relapses in large phase III studies: fingolimod was approved in September 2010, and cladribine was submitted to the FDA for approval for use in MS, but further safety studies were requested.

Fingolimod

Fingolimod is a relatively small lipophilic chemical (Fig. 7.4) derived from fungi, related to the myriocins, which have antibiotic properties. The fungus source, *Isaria sinclairii*, has been used as a Chinese herbal remedy for centuries. The drug is thought to work by interfering with sphingosine-1-phosphate (S1P) binding to its receptors in lymph nodes further discussed in Chap. 19.

Cladribine

This drug is a small purine analog (molecular weight of 286) which resembles adenosine and inhibits adenosine deaminase. It has been used for the last 20 years in the treatment of hairy cell leukemia and is considered an immunosuppressive agent. Positive treatment results in MS were reported in the early 2010 [22]. An oral treatment protocol in the phase 3 study results in prolonged leukopenia. Three to four percent of patients developed shingles (reactivated herpes zoster) and one patient died of reactivated latent tuberculosis. Otherwise the drug appeared to be well-tolerated.

1.5.6   Combination Therapies

Not only is the pathogenesis of MS complex but also factors contributing to disease may vary among patients. This would suggest that combination therapy strategies might be effective. Thus, some clinicians will initially use the DMDs, and add to this corticosteroids, azathioprine, natalizumab, cyclophosphamide, or methotrexate. Although combination therapies are frequently used in practice, they have not been demonstrated to be effective in large clinical studies [23].

2   Symptomatic Therapies

A neurologist caring for MS patients treats not only the disease itself but also a variety of symptoms caused by the CNS damage. The most common symptoms are fatigue, bladder and bowel dysfunction, spasticity, cognitive problems, depression, and sexual problems. There are two categories of approach to therapy: life-style changes and rehabilitation, and medications. This topic has been recently reviewed in depth [24].

(a) *Rehabilitation.* Rehabilitation medicine has proven extremely helpful to MS patients and should be utilized extensively. Although traditionally treating patients with gait difficulty, rehabilitation medicine physicians and physical therapists have broadened their scope to provide a range of other services to MS patients.

(b) *Life-style changes.* In early MS, when disability is mild, patients can usually pursue their lives with little attention to the disease. But when disability begins to develop, certain life-style changes can be very effective. For instance, rest during the day and careful attention to improving sleep can help fatigue. Bladder symptoms can be improved by cutting back on commonly ingested diuretics such as caffeine.

(c) *Medications.* A wide range of medications can be effective for MS symptoms. However, many of these medications have significant side effects and a careful risk/benefit analysis must be performed with each patient. For instance, anti-spasticity medications such as baclofen and tizanidine can cause a generalized lowering of muscle tone which can interfere with gait; they also frequently increase fatigue.

Spasticity, the increased muscle and tendon tone, and stiffness of movement of weak muscles associated with CNS damage, can be disabling to patients with MS and other diseases which result in CNS injury. Four different classes of agents are used to decrease spasticity in multiple sclerosis: (a) GABA agonists (e.g., baclofen), (b) α-2 adrenergic agonists (e.g., tizanidine), (c) peripheral anti-spastics (e.g., dantrolene), and (d) botulinum toxin injected locally to decrease spasticity. Of these the most widely prescribed medication for spasticity in MS is baclofen (brand name in the USA: Lioresal). The rationale for the use of baclofen is as follows: neurons of the cerebral cortex have been classified into projection neurons, which primarily use glutamate as an excitatory neurotransmitter, and local interneurons,

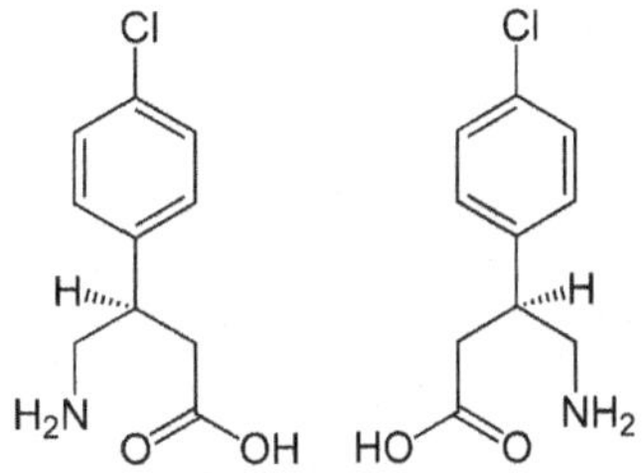

Fig. 7.5 Chemical structure of baclofen

which use GABA as an inhibitory neurotransmitter. Many GABA-ergic interneurons also used a variety of other neurotransmitter such as opiate peptides, cholecystokinin, and somatostatin. GABA-ergic interneurons in the spinal cord also frequently use glycine as a neurotransmitter.

The structure of baclofen is interesting; it is a small molecule (molecular weight of 214), composed of two enantiomers, mirror images of each other, in a 1:1 ratio (Fig. 7.5). It is considered a GABA agonist, and in this way substitutes for the GABA produced in the spinal cord by inhibitory interneurons. The GABA receptors are located in the presynaptic terminals of the Ia fibers, binding of GABA or baclofen to these receptors inhibits calcium influx and lowers the amount of transmitter released.

Patients with progressive MS can develop spasticity in their arms and hands which can sometimes be relieved by local injections of botulinum toxin (Inset 7.1). Botulinum toxin can also be injected into the detrusor muscle of the bladder which can improve bladder symptoms in some MS patients. Bladder issues are especially common in MS, and, depending on their cause can be improved with anticholinergics or antibiotics for urinary infections. An increasingly helpful referral for some patients is to neuro-urologists, who can provide the latest and best diagnosis and treatments for MS patients with difficult-to-manage bladder control problems.

A medication that has recently been approved as a symptomatic medication for improving the gait of patients with MS is 4-aminopyridine (Ampyra), also called dalfampridine or fampridine. Historically, 4-AP has been used in the research laboratory to characterize various forms of the potassium channel, since it is a relatively

Inset 7.1 Botulinum Toxin

The word botulinum comes from the Latin word for sausage, botulus, since in the nineteenth century botulism, i.e., poisoning with botulinum toxin, was most commonly seen in its food-borne form, from improperly handled meats as ingredients for sausages. The toxin is produced by the anaerobic bacterium *Clostridium botulinum*, a Gram-positive rod; the pathogen was first isolated by Emile van Ermengem, a Belgian bacteriologist in 1896 from a cured ham identified as the cause of a botulism outbreak.

Botulinum toxin is a large protein (molecular weight of 150,000) made up of two chains, one of which is a protease that acts on a fusion protein at the presynaptic terminal of the neuromuscular junction to inhibit acetylcholine release (the neuromuscular junction as an important "Achilles heel" in neuroimmunology is discussed in Chap. 10 on myasthenia gravis). The lethal dose for inhaled botulinum toxin is about 3 ng/kg.

Botulinum toxin poisoning is important to neurologists as a cause of rapidly progressive paralysis, and as an important differential diagnosis in patients with Guillain–Barré syndrome (see Chap. 9). Botulinum toxin is available commercially for therapeutic purposes which is especially important to neurologists as an aid in the treatment of spasticity, especially when a muscle or small group of muscles can be targeted. Many patients with chronic myelopathies, including MS, have clenched hands that can be loosened up by botulinism toxin injections. Other uses of botulinum toxin are cosmetic or military, as part of many countries' arsenal of biological weapons.

selective blocker of a family of voltage-gated potassium channels. In MS, there are theoretically augmented potassium currents which ultimately can lead to conduction failure. Blocking potassium currents with 4-AP can reverse this effect, and the drug has been used in MS and the Lambert–Eaton myasthenic syndrome (see Chap. 9). The action of this drug, however, may not be so simply explained, since recent studies have shown that 4-AP also has effects on the calcium channel.

References

1. Ebers GC, Heigenhauser L, Daumer M, Lederer C, Noseworthy JH. Disability as an outcome in MS clinical trials. Neurology. 2008;71(9):624–31.
2. Cohen JA, Barkhof F, Comi G, et al. Oral fingolimod or intramuscular interferon for relapsing multiple sclerosis. N Engl J Med. 2010;362(5):402–15.
3. Jacobs LD, Beck RW, Simon JH, et al. Intramuscular interferon beta-1a therapy initiated during a first demyelinating event in multiple sclerosis. CHAMPS Study Group. N Engl J Med. 2000;343(13):898–904.
4. Confavreux C, Vukusic S, Moreau T, Adeleine P. Relapses and progression of disability in multiple sclerosis. N Engl J Med. 2000;343(20):1430–8.
5. Scalfari A, Neuhaus A, Degenhardt A, et al. The natural history of multiple sclerosis, a geographically based study 10: relapses and long-term disability. Brain. 2010;133(Pt 7):1914–29.
6. Beck RW, Cleary PA, Anderson Jr MM, et al. A randomized, controlled trial of corticosteroids in the treatment of acute optic neuritis. The Optic Neuritis Study Group. N Engl J Med. 1992;326(9):581–8.
7. Pachner AR, Warth JD, Pace A, Goelz S. Effect of neutralizing antibodies on biomarker responses to interferon beta: the INSIGHT study. Neurology. 2009;73(18):1493–500.
8. Filippini G, Munari L, Incorvaia B, et al. Interferons in relapsing remitting multiple sclerosis: a systematic review. Lancet. 2003;361(9357):545–52.
9. Pachner AR. An improved ELISA for screening for neutralizing anti-IFN-beta antibodies in MS patients. Neurology. 2003;61(10):1444–6.
10. Pachner AR, Steiner I. The multiple sclerosis severity score (MSSS) predicts disease severity over time. J Neurol Sci. 2009;278(1–2):66–70.
11. Pachner AR, Dail D, Pak E, Narayan K. The importance of measuring IFNbeta bioactivity: monitoring in MS patients and the effect of anti-IFNbeta antibodies. J Neuroimmunol. 2005;166(1–2):180–8.
12. Teitelbaum D, Meshorer A, Hirshfeld T, Arnon R, Sela M. Suppression of experimental allergic encephalomyelitis by a synthetic polypeptide. Eur J Immunol. 1971;1(4):242–8.
13. Karussis D, Teitelbaum D, Sicsic C, Brenner T. Long-term treatment of multiple sclerosis with glatiramer acetate: natural history of the subtypes of anti-glatiramer acetate antibodies and their correlation with clinical efficacy. J Neuroimmunol. 2010;220(1–2):125–30.

14. Brenner T, Arnon R, Sela M, et al. Humoral and cellular immune responses to Copolymer 1 in multiple sclerosis patients treated with Copaxone. J Neuroimmunol. 2001;115(1–2):152–60.

15. Rubio-Terres C, Dominguez-Gil Hurle A. [Cost-utility analysis of relapsing-remitting multiple sclerosis treatment with azathioprine or interferon beta in Spain]. Rev Neurol. 2005;40(12):705–10.

16. Casetta I, Iuliano G, Filippini G. Azathioprine for multiple sclerosis. Cochrane Database Syst Rev. 2007; (4):CD003982.

17. Ransohoff RM. Natalizumab for multiple sclerosis. N Engl J Med. 2007;356(25):2622–9.

18. Edan G, Miller D, Clanet M, et al. Therapeutic effect of mitoxantrone combined with methylprednisolone in multiple sclerosis: a randomised multicentre study of active disease using MRI and clinical criteria. J Neurol Neurosurg Psychiatry. 1997;62(2):112–8.

19. Hartung HP, Gonsette R, Konig N, et al. Mitoxantrone in progressive multiple sclerosis: a placebo-controlled, double-blind, randomised, multicentre trial. Lancet. 2002;360(9350):2018–25.

20. Marriott JJ, Miyasaki JM, Gronseth G, O'Connor PW. Evidence Report: The efficacy and safety of mitoxantrone (Novantrone) in the treatment of multiple sclerosis: report of the Therapeutics and Technology Assessment Subcommittee of the American Academy of Neurology. Neurology. 2010;74(18):1463–70.

21. Boster A, Edan G, Frohman E, et al. Intense immunosuppression in patients with rapidly worsening multiple sclerosis: treatment guidelines for the clinician. Lancet Neurol. 2008;7(2):173–83.

22. Giovannoni G, Comi G, Cook S, et al. A placebo-controlled trial of oral cladribine for relapsing multiple sclerosis. N Engl J Med. 2010;362(5):416–26.

23. Conway D, Cohen JA. Combination therapy in multiple sclerosis. Lancet Neurol. 2010;9(3):299–308.

24. Kesselring J, Beer S. Symptomatic therapy and neurorehabilitation in multiple sclerosis. Lancet Neurol. 2005;4(10):643–52.

Experimental Models of MS

8

Multiple sclerosis is a uniquely human disease. It does not occur naturally in any other animal, and, in fact, no other disease closely resembling MS has been described in other animals. Similarly, no animal models of MS mimic all of the features of the human disease faithfully. However, animal models of MS are important because they allow us to learn about neuroinflammation, and its link to demyelination, neuronal injury, and disability. The two main types of models—autoimmune and viral—also teach us about the effects of these processes within the CNS. We learn something from each model, and it is important to assess information obtained from each one as contributing to our knowledge base, rather than having direct translation to the human disease. This point is frequently lost on individuals who wish to rush preliminary information about models directly to MS treatments. This eagerness is understandable in MS patients, who are desperate for more effective therapies. But clinicians and scientists must take the long view, and place new information in context of the strengths and weaknesses of each model.

There are many problems with modeling MS. First, the etiopathogenesis of the human disease is unknown. The two main types of models, autoimmunity and virus, are based on reasonable hypotheses, but there is no strong evidence that MS is due to either an autoimmune or a viral process and in fact, there is no assurance that the underlying cause is the same in all patients. Second, MS is multifaceted, involving attacks, progressive disability, demyelination, inflammation, and excessive immunoglobulin production

within the CNS. Specific models can focus on only a few of these aspects of MS at a time, but none reproduces the whole spectrum of the disease. Third, the ideal use of an animal model is to provide information about basic biological processes within the CNS with some relevance to MS, but increasingly there is pressure from patients and funding agencies to translate work in animal models to human disease too rapidly. Despite these and other problems, the use of animal models for MS has continued to be used extensively, and indeed provides the most common research focus within neuroimmunology.

1 Animal Models of MS

1.1 Autoimmune

> "EAE, the autoimmunity-based animal model, is not seen as inherently superior to (viral) models like Visna or TMEV…One might be right to regard autoimmunity as a paradigm shift that never quite made it!" Byron H. Waksman [1]

Under certain conditions provoking an autoimmune response in the CNS, animals can develop inflammation, demyelination, and weakness. The first demonstration that the immune response in presumably healthy individuals could injure the CNS came when humans, injected with rabies vaccines containing CNS material to prevent rabies infections, developed post-vaccinial encephalomyelitis, reviewed in Chap. 3. In 1925, Koritschoner and Scweinburg induced inflammation in rabbit

A.R. Pachner, *A Primer of Neuroimmunological Disease*, DOI 10.1007/978-1-4614-2188-7_8, © Springer Science+Business Media, LLC 2012 spinal cord by immunization with human spinal cord [2]; soon thereafter, Rivers et al. published a report of their work on reproducing systematically this syndrome in monkeys, also called acute disseminated encephalomyelitis (ADEM), by immunization with CNS tissue [3]. Since then, experimental autoimmune encephalomyelitis (EAE) has been induced in many different experimental animals by immunization with any one of several different CNS antigens. Although initially EAE was developed to model ADEM, the combination of inflammation and demyelination led investigators to recognize that it also mimic features of MS.

CNS tissue, or purified parts of it, that can induce EAE are called encephalitogens. Initial work on EAE was focused on determining the composition of encephalitogens [4]. As immunology boomed in the second half of the twentieth century, and the research on encephalitogens resulted in reproducibility of the model, the immunology of EAE attracted more and more interest, especially as a rapidly induced model of autoimmunity. Most EAE projects currently utilize mice and rats, but in the past, rabbits, guinea pigs, nonhuman primates, and other animals were used and continue to be used by some groups. The most productive non-rodent work over the last 15 years has been in marmoset monkeys [5, 6]. Work on monkeys always has greater relevance to human disease because of the proximity of these animals to humans on the phylogenetic tree relative to rodents. The pathology of the marmoset EAE model appears to more closely resemble MS than rodent EAE, but nonhuman primates are much more difficult to work with than are rodents.

Encephalitogens currently used for EAE research vary, and some laboratories still use spinal cord or brain homogenates or myelin. However, most laboratories use more purified encephalitogens such as myelin basic protein (MBP), proteolipid protein (PLP), or myelin oligodendrocyte protein (MOG), or synthetic peptides which represent part of the sequences of these proteins. The most commonly used strain of mouse for these experiments is the SJL strain, which appears to have poorly defined genetically determined traits

which make it more susceptible than other strains. In MOG-induced EAE, a different mouse strain, C57Bl/6, is more susceptible. Most investigators utilize active immunization of the encephalitogen combined with an adjuvant, usually Freund's adjuvant, a mixture of oil and other constituents, including heat-killed mycobacteria, that enhances the immune response to the encephalitogen, possibly by the stimulation of toll-like receptors (TLRs) . In addition, most murine EAE models require the injection of a bacterial toxin, pertussis toxin (see Inset 8.1), which is thought to enhance EAE by increasing blood–brain barrier permeability or by activating TLRs.

Inset 8.1

Pertussis toxin is a large molecule produced by the bacterium, *Bordetella pertussis*, composed of six subunits and is an exotoxin, meaning it is secreted by the bacterium. An effect that is important for neuroimmunologists is that it is considered essential in many forms of EAE. It is thought to exert this effect by impairing the blood–brain barrier and facilitating the influx of lymphocytes into the CNS. The precise mechanism by which the toxin causes this effect is unknown, but it may have a direct effect on CNS endothelial cells in experimental animals. *Bordetella pertussis* is the causative pathogen of the severe childhood respiratory infection, whooping cough.

The natural course of most forms of EAE is the development of weakness within about a week after immunization, with disability peaking at about 10–14 days after immunization, followed by resolution to baseline strength by 3–4 weeks after immunization. The demyelination and inflammation in the CNS usually follows this temporal pattern also. The weakness in EAE is associated with flaccidity, i.e., decreased muscle tone, a feature of EAE in which it differs from

MS, where spasticity, i.e., increased muscle tone, caused by CNS injury, is predominant. The flaccidity found in EAE may be due to the fact that many EAE models have nerve or nerve root involvement. Most EAE investigators use a relatively subjective visual assessment grading scale of zero to five plus for grading weakness rather than objective neurobehavioral analyses. There are many variations on this model, since mutant mice (see Inset 8.2) can be genetically engineered to overexpress or underexpress a wide variety of genes which have effects on the course of EAE.

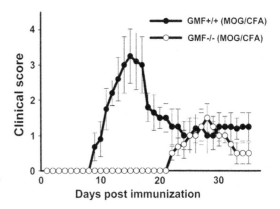

Fig. 8.1 Course of MOG-induced EAE in mice. The x-axis is days post-immunization with a myelin oligodendrocyte glycoprotein (MOG) peptide and the y-axis is a clinical score based on visualization of the mouse and estimate of its weakness. EAE is suppressed in glial maturation factor (GMF) knockout mice. The wild-type, i.e. normal mice, show normal development of EAE with peak 14 days after immunization with MOG peptide

Inset 8.2 Mutant Mice

Genetically engineered mice, also called mutant mice, have been used extensively in neuroimmunology research, especially in models of EAE. Mutant mice can be loosely divided into knockout mice which have a deletion of a gene, and transgenic mice, in which a gene is inserted. For instance, a recently developed mutant mouse is the MOG peptide 92–106 T-cell receptor transgenic mouse in which most T cells express this genetically engineered receptor and react to MOG [21]; this mouse spontaneously gets EAE without the need for immunization or other manipulations.

In 2007, Capecchi, Smithies, and Evans received the Nobel Prize for the development of knockout mice, utilizing homologous recombination in which nucleotide sequences are exchanged between similar strands of DNA. The first knockout mouse was made in 1989. An example of the use of knockout mice in EAE research is the observation that EAE can be induced in mice in whom the components critical for the membrane attack complex (MAC), an important part of the complement system, have been knocked out [22]. These experiments indicate that MAC is not an essential set of genes in EAE induction; it would have almost impossible to prove this if a mutant mouse was not available.

The temporal pattern of EAE (Fig. 8.1) is essentially identical to that of post-vaccinial autoimmune encephalomyelitis, as outlined in Chap. 3, for the Semple rabies vaccine encephalomyelitis (SAE). Stage 1 is the immunization with the encephalitogen. In contrast to SAE, in which the main reason for the immunization is protection from rabies, and the sensitization to CNS antigens is an unwanted side effect, in EAE, the sensitization to the encephalitogen is the main purpose of the immunization. In stage 2, usually in the first few days after the immunization, local lymph nodes drain the site of immunization, and the antigen is processed within these nodes by dendritic cells and other antigen-processing cells such as macrophages. By the end of the first week, in stage 3, CNS antigen-specific lymphocytes proliferate and enter the circulation, and begin to enter the CNS. Once they enter the CNS encephalitogen-specific lymphocytes are retained there by the presence of the antigen and do not traffic out of the CNS, while lymphocytes with specificities for other antigens are not retained in the CNS, and either traffic back into the circulation, or alternatively apoptose within the CNS (see apoptosis, Chap. 1), i.e., undergo programmed cell death (PCD).

Stage 4, spanning from about day 10 to day 20 after immunization, represents the period of most active inflammation within the CNS and also is the period of the development of maximum neurological signs in the animal. The animal's spinal cord, which under normal circumstances contains only an occasional mononuclear cell, will have sites of large numbers of them, mostly lymphocytes and macrophages. Initially, mononuclear cells accumulate in the perivascular space between the basement membranes of the cerebrovascular endothelial cells and of the basement membrane associated with astrocytic endfeet (Fig. 3.1). Over time, the mononuclear cells move through the basement membrane and the glia limitans comprising astrocytic endfeet and into the parenchyma of the CNS. Demyelination also appears at this stage, although the extent of demyelination varies widely in different models of EAE.

By 3 weeks after immunization in stage 5, inflammation is abating and recovery is beginning. EAE, in most models, is monophasic and after the peak of inflammation and disease animals improve relatively rapidly, often to being completely normal. In some EAE models, the animals improve markedly after the peak of disease, but do not completely recover back to their baseline and there is permanent injury to the CNS caused by the acute phase of the disease. Finally, some models are described as being "chronic, relapsing EAE" [7]. However, the degree of chronicity is usually mild. For instance, in a recent publication authors evaluated a therapy, surgical lymph node excision, in three different models of mouse EAE, which they described as "acute, chronic, and chronic relapsing", induced by immunizing SJL, C57Bl/6, or Biozzi ABH mice with proteolipid protein peptide, myelin oligodendrocyte glycoprotein peptide (35–55), or myelin oligodendrocyte glycoprotein (8–21) [8]. However, although there were minor differences in the course of the neurological disability in these mice, in all three groups of animals the disability peaked at 10–15 days post-immunization and had completely resolved by the end of the second month. Thus, EAE is generally a mono-

phasic phenomenon, and, given the absence of accumulating neurological injury, it is not a good model for the progressive disability of MS, that aspect of the disease most important to the patient and their treating physician.

Groundbreaking work by Ben-Nun, Wekerle, and others in the early 1980s [9] revealed that T cells could mediate EAE, which led to hope that a new era of understanding EAE and possibly MS was dawning. Since then the T cell has been the primary cell researched in EAE. In fact, because of the importance of the T cell in EAE, many investigators extrapolated to MS by assuming that MS was a "T-cell-mediated autoimmune disease." However, it appears that the injury to the CNS in most forms of EAE is caused by a complex mix of components that cannot be reproduced by a single cell type. For instance, anti-MOG antibody is necessary for maximum clinical disease in some models of MOG-induced EAE, and weakness can be amplified by the administration of anti-MOG monoclonal antibodies, but not antibodies to irrelevant antigens.

The precise type of T cell or mixture of T cells which are most pathogenic in EAE is still a matter of controversy. In Ben-Nun's original manuscript describing T-cell mediation of EAE [9], lymphocyte lines were utilized. These were complex mixtures of cells, not T cell clones, and since that pivotal observation, populations of T cells, presumably interacting together, rather than a monoclonal population, have been the most potent effectors. Most investigators until recently assumed that the most pathogenic cells were CD4+, interferon-γ secreting cells (sometimes called Th1 cells). However, recent evidence points to the importance of Th17 cells, CD4+ T cells that secrete IL-17, which are developmentally distinct from Th1 and Th2 cells (Fig. 8.2). In humans, these cells are induced by a combination of IL-23, IL-1β, and TGF-β and secrete IL-17 and IL-22.

There are many variants of EAE, some of them made possible by the availability of genetically altered mutant mice. A notable one is the TCR-transgenic mouse [10] in which a large percentage

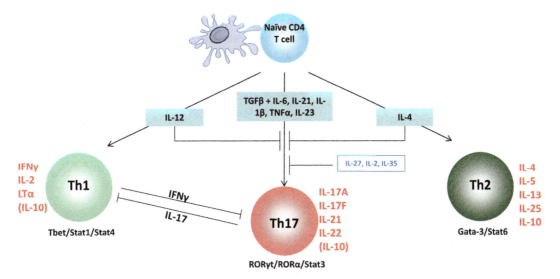

Fig. 8.2 Differentiation of effector T helper subsets from naïve CD4 T cells. The first step is activation of naïve CD4+ T cells by professional antigen-presenting cells (APCs) with IL-12 being an important stimulatory cytokine. Th1 cells upregulate IFN-γ, while Th2 cells dif- ferentiate in response to IL-4. The Th17 cell subset develops in response to IL-6 or IL-21 and TGF-β, which is enhanced in the presence of IL-1β and TNF-α. Thus, T-cell differentiation is determined by a combination of cell type and cytokine environment

of T-cell receptors are specific for MBP. These mice develop EAE simply with the administration of pertussis toxin. One way of producing EAE in an animal is passive transfer of cell populations into naive mice, which allows determination of which types of cells are critical for EAE induction (see "passive transfer", below). Another interesting recently developed model is that of focal lesions causing demyelination, inflammation, and weakness, induced by intraspinal injection of the protein vascular endothelial growth factor (VEGF) in rats with high levels of anti-MOG antibodies [11]. These variations of the "standard" EAE provide greater opportunities to learn about the biology of EAE.

Because of the many dissimilarities between EAE and MS, a direct extrapolation of findings in EAE to MS is unwarranted. A recent editorial, titled "Experimental Allergic Encephalomyelitis: A Misleading Model of Multiple Sclerosis" [12], concluded that "we therefore are forced to examine MS without the restraints of EAE", and makes a reasonable argument that EAE has been inappropriately used to provide a framework for

attempts to understand MS. Ultimately, EAE can be considered important as a means to understanding inflammation in the CNS. Despite its limitations, much will continue to be learned from EAE about a wide variety of phenomena relevant to human disease including the blood–brain barrier, effects of resident CNS populations, immune cell trafficking through the CNS, and cytokine production and cytokine effects.

1.2 Viral

CNS demyelination is produced by a wide variety of viruses in animals [13]; the list includes visna virus infection in Icelandic sheep, caprine arthritis-encephalitis virus infection in goats, Semliki Forest virus in mice, and canine distemper and its variants in dogs and sea mammals. These viruses represent a broad spectrum of viruses without any similarities, except that they all are RNA viruses. The most widely used viral models for MS are the Theiler's murine encephalomyelitis virus (TMEV) and mouse hepatitis virus (MHV) models.

TMEV is a picorna virus, a very small virus of only 8,100 nucleotides in the Cardiovirus genus. After intracerebral injection of the virus into a mouse, there are two phases of the disease. The first is an acute mild, usually subclinical encephalitis, but if too much virus is injected or if the mouse is immunosuppressed, the early encephalitis can be deadly. During this first phase, the infection is predominantly of neurons. The chronic phase beginning about a month after infection consists of slowly progressive disability, characterized by demyelination, inflammation, and axonal damage. The weakness is associated with spasticity and occasionally severe muscle spasms. The viral load in the CNS in the chronic phase is stable over time, with the primary reservoir being astrocytes, and the progressive damage to the CNS is presumed to be due to the large amount of inflammation in the CNS. Except for the lack of relapses the clinical picture resembles human MS; the pathology is very similar also, as is the presence of plasma cells and production of IgG within the CNS. A disadvantage for researchers, though, is that mice can take months to become weak after the initial infection; in contrast, weakness in EAE occurs within a few weeks of immunization.

The TMEV model has a number of advantages over EAE as outlined in the table below. Neither TMEV nor EAE mirror the combination of exacerbations/remissions and progressive disability of the human disease. However, the progressive disability of the TMEV model is particularly attractive as a target since neurologists, at this time in the history of multiple sclerosis, have so many therapies that have been shown to be effective for exacerbations (see Chap. 7), and what is needed are therapies that affect disability progression. For reasons that are not clear in either MS or TMEV, both diseases progress through 20–25% of the life span of the animal and then usually almost completely plateau; i.e., neither patients nor mice commonly die of progressive neurological disability. Both TMEV and MS have prominent involvement of the humoral arm of the immune response, with antibody being both produced and deposited in the CNS; in contrast, in most models of EAE, there is little evidence of B-cell or antibody involvement.

Characteristic	MS	TMEV	EAE
Exacerbations/remissions and progressive disability	Yes	No	No
Progressive disability	Yes	Yes	No
Monophasic, hyperacute disease with rapid recovery	No	No	Yes
Viral etiology	?	Yes	No
Inflammation	Yes	Yes	Yes
Demyelination	Yes	Yes	Yes
Disability progression over what % of life span	25%	20%	No progression
Antibody production in the CNS	Yes	Yes	No
Immunoglobulin deposition in the CNS	Yes	Yes	No

In the MHV model, also called JHMV from the most commonly used strain of MHV, demyelination, inflammation, and high viral levels in the CNS occur as a monophasic event, followed by slow diminution of viral load and partial resolution of demyelination and inflammation. However, the virus is not completely cleared and new areas of demyelination appear for prolonged periods, possibly due to local reactivation of virus in areas of the CNS [13]. MHV is also a well-studied model of acute viral encephalitis (see Chap. 12).

Canine distemper virus (CDV) and other related viruses are morbilliviruses related to measles that cause demyelination in a variety of animals in the wild; these viruses are the subjects of some recent reviews [14, 15]. The most medically important morbillivirus is measles virus which generally causes an acute viral illness in children, and has a significant mortality especially in the developing world where measles vaccination is less common. Measles virus, however, also causes two rare syndromes, subacute sclerosing panencephalitis (SSPE) and measles inclusion body encephalitis (MIBE), described in more depth in Chap. 16; SSPE can mimic MS.

CDV, which in the wild has primarily affected dogs and also wild cats and other mammals, initially causes a systemic infection, transmitted through the lungs, that ultimately invades the CNS, and results in glial cell infection. Subsequently, the virus persists chronically and causes multifocal, inflammatory demyelinating lesions resembling MS. The precise pathogenesis of the lesions is controversial, but appears to be related to a combination of persistent virus, macrophage activation, strongly upregulated cytokine responses, and an inflammatory pathology consisting of perivascular infiltration with plasma cells and lymphocytes. Oligodendrocytes, though infected by CDV, are not lysed. Morbilliviruses also cause inflammatory demyelination in marine mammals including seals, dolphins, whales, and porpoises.

Retrovirus infections can also cause demyelinating, inflammatory disease in animals; the most well-known retrovirus causing human disease is human immunodeficiency virus (HIV), the neurological manifestations of which will be reviewed in Chap. 12. Visna virus was the first retrovirus isolated in 1957 and is similar to TMEV in causing a subacute encephalitis, followed by a chronic demyelinating encephalomyelitis; susceptible animals are sheep. Another retrovirus causing demyelination is caprine arthritis encephalitis virus (CAEV), which causes inflammatory demyelination in the brain and spinal cord in young goats.

1.3 Demyelination Induced by Toxins

Demyelination can be induced also by administration of toxic agents in noninflammatory models of the demyelination of MS. In the cuprizone model of CNS demyelination, mice are fed cuprizone, a copper chelator, and subsequently develop severe demyelination due to cell death of oligodendrocytes. Although the effect is thought to be due to the copper chelation, the effect is not reversed by supplementation with copper. Once the cuprizone is removed, remyelination begins to occur, so this model can also be used to look at processes which might hinder or accelerate remyelination. If focal demyelination is required, that can be induced by injection with lysolecithin or related compounds. These and similar models have been recently reviewed [16].

2 In Vitro Models of MS

Some researchers utilize isolated CNS tissue, such as brain slices or isolated optic nerves, to investigate the role of various cells or molecules to injure or demyelinate. For instance, cerebrospinal fluid from MS patients but not controls has been shown to block transmission of action potentials through isolated optic nerves. Other researchers have used isolated optic nerves from animals with EAE to analyze the cytokines and mediators involved in the immune attack on myelin. One advantage of these ex vivo models is that functional aspects of axonal conduction such as action potentials can be correlated with measurement of immune cells and molecules.

Advances in the culturing of cells ex vivo has allowed studies of CNS endothelial cell function, commonly using inserts in which endothelial cells are grown on a porous substrate that allows molecules and cells to pass from one chamber to another, mimicking the blood–brain barrier. Since increased passage of mononuclear cells across CNS endothelial cells is an important pathology in EAE and MS, this technology has yielded important findings. For instance, the cell culture insert approach was helpful in the analysis of agents such as natalizumab which target adhesion molecules expressed on endothelial cells.

3 Animal Models of Neuromyelitis Optica (NMO) and the Use of Passive Transfer Models

NMO, described in Chap. 6, is an important disease for at least three reasons: first, it is frequently a particularly devastating illness causing progressive visual loss and spinal cord injury; second, it is frequently misdiagnosed as MS and needs to be differentiated from that disease because treatment is different; and third, its etiopathogenesis is much better defined than MS in that it appears to be caused by an autoantibody. For all these reasons, there has been extensive interest in developing animal models, and a number of laboratories have been successful [17, 18]. In these models, the mediator of injury is IgG from patients with NMO which results in injury either after being directly injected into the CNS or via intravenous injection after inducing EAE and opening the blood–brain barrier. The pathogenic IgG can be either isolated directly from the blood of NMO patients or can be made by recombinant technology. Thus, these models fall into a category of experimental models called passive transfer in which the disease can be induced in a normal animal by injection of pathogenic immunoglobulin or cells.

The definition of passive transfer is the conferring of an immune response to a nonimmune host by injection of immunoglobulin or lymphocytes from an immune donor; thus, passive transfer models of neuroimmunological disease involve the transfer of cells or immunoglobulin into an animal to induce a model of disease. The most commonly used model using passive transfer of IgG in neuroimmunology is that of passively transferred myasthenia gravis (MG), first used by Toyka [19]; experimental models of MG are described in Chap. 10. Passive transfer EAE, using mononuclear cells from immune animals, has also been utilized extensively, allowing characterization of pathogenic cell populations. Passive transfer EAE can also be elicited with anti-MOG antibodies [20]. Passive transfer experiments in a variety of diseases such as EAE, experimental myasthenia, and experimental NMO have proved helpful in identifying the populations of cells or antibodies which might be pathogenic in human disease.

For those interested in a more detailed insight into various aspects of experimental models of MS, I recommend the book, Experimental models of multiple sclerosis, edited by Ehud Lavi, Cris S. Constantinescu, Springer, 2005.

References

1. Waksman BH. Demyelinating disease: evolution of a paradigm. Neurochem Res. Apr 1999;24(4):491–5.
2. Gold R, Linington C, Lassmann H. Understanding pathogenesis and therapy of multiple sclerosis via animal models: 70 years of merits and culprits in experimental autoimmune encephalomyelitis research. Brain. Aug 2006;129(Pt 8):1953–71.
3. Rivers TM, Spunt DH, Berry GP. Observations on attempts to produce acute disseminated encephalomyelitis in monkeys. J Exp Med. 1933;58:39–53.
4. Waksman BH, Porter H, Lees MD, Adams RD, Folch J. A study of the chemical nature of components of bovine white matter effective in producing allergic encephalomyelitis in the rabbit. J Exp Med. 1954;100(5):451–71.
5. t Hart BA, Laman JD, Bauer J, Blezer E, van Kooyk Y, Hintzen RQ. Modelling of multiple sclerosis: lessons learned in a non-human primate. Lancet Neurol. Oct 2004;3(10):588–97.
6. Genain CP, Hauser SL. Experimental allergic encephalomyelitis in the New World monkey Callithrix jacchus. Immunol Rev. Oct 2001;183:159–72.
7. Steinman L. Assessment of animal models for MS and demyelinating disease in the design of rational therapy. Neuron. Nov 1999;24(3):511–4.
8. van Zwam M, Huizinga R, Heijmans N, et al. Surgical excision of CNS-draining lymph nodes reduces relapse severity in chronic-relapsing experimental autoimmune encephalomyelitis. J Pathol. Mar 2009; 217(4):543–51.
9. Ben-Nun A, Wekerle H, Cohen IR. The rapid isolation of clonable antigen-specific T lymphocyte lines capable of mediating autoimmune encephalomyelitis. Eur J Immunol. Mar 1981;11(3):195–9.
10. Goverman J, Woods A, Larson L, Weiner LP, Hood L, Zaller DM. Transgenic mice that express a myelin basic protein-specific T cell receptor develop spontaneous autoimmunity. Cell. 1993;72(4): 551–60.
11. Sasaki M, Lankford KL, Brown RJ, Ruddle NH, Kocsis JD. Focal experimental autoimmune encephalomyelitis in the lewis rat induced by immunization with myelin oligodendrocyte glycoprotein and intraspinal injection of vascular endothelial growth factor. Glia. 2010;58:1523–31.
12. Sriram S, Steiner I. Experimental allergic encephalomyelitis: a misleading model of multiple sclerosis. Ann Neurol. Dec 2005;58(6):939–45.
13. Stohlman SA, Hinton DR. Viral induced demyelination. Brain Pathol. Jan 2001;11(1):92–106.
14. Vandevelde M, Zurbriggen A. Demyelination in canine distemper virus infection: a review. Acta Neuropathol. Jan 2005;109(1):56–68.
15. Sips GJ, Chesik D, Glazenburg L, Wilschut J, De Keyser J, Wilczak N. Involvement of morbilliviruses in the pathogenesis of demyelinating disease. Rev Med Virol. 2007;17(4):223–44.
16. Blakemore WF, Franklin RJ. Remyelination in experimental models of toxin-induced demyelination. Curr Top Microbiol Immunol. 2008;318:193–212.
17. Bennett JL, Lam C, Kalluri SR, et al. Intrathecal pathogenic anti-aquaporin-4 antibodies in early neuromyelitis optica. Ann Neurol. Nov 2009;66(5): 617–29.

18. Bradl M, Misu T, Takahashi T, et al. Neuromyelitis optica: pathogenicity of patient immunoglobulin in vivo. Ann Neurol. Nov 2009;66(5):630–43.

19. Toyka KV, Drachman DB, Griffin DE, et al. Myasthenia gravis. Study of humoral immune mechanisms by passive transfer to mice. N Engl J Med. 1977; 296(3):125–31.

20. Lyons JA, Ramsbottom MJ, Cross AH. Critical role of antigen-specific antibody in experimental autoimmune encephalomyelitis induced by recombinant myelin oligodendrocyte glycoprotein. Eur J Immunol. 2002;32(7):1905–13.

21. Pollinger B, Krishnamoorthy G, Berer K, et al. Spontaneous relapsing-remitting EAE in the SJL/J mouse: MOG-reactive transgenic T cells recruit endogenous MOG-specific B cells. J Exp Med. 2009;206(6):1303–16.

22. Barnum SR, Szalai AJ. Complement and demyelinating disease: no MAC needed? Brain Res Rev. 2006; 52(1):58–68.

Guillain–Barré Syndrome (GBS) and Other Immune-Mediated Neuropathies

1 Definition

Guillain–Barré syndrome (GBS) is the most common serious immune-mediated disease affecting the peripheral nervous system (PNS). GBS is an acute diffuse illness affecting nerve roots and peripheral nerves, and occasionally the cranial nerves, often ascending from the feet over hours to days to the arms and sometimes bulbar muscles, resulting in motor weakness, and relatively sparing sensory functions; most patients who survive the acute paralysis recover completely within months. The disease was first described by the French neurologist, Landry, in 1859 (Inset 9.1).

Inset 9.1 What Should This Disease Be Called?

Most physicians call this disease as Guillain–Barré syndrome (GBS), although it was originally described by Jean Landry de Thézillat (usually referred to as Landry) in 1859 before Georges Guillain or Jean Alexander Barré was born; thus the disease is sometimes called Landry's paralysis. Despite the fact that Andre Strohl, a physiologist, contributed to Guillain and Barré's work, his name is not frequently added to the syndrome. Thus, if one were a strict eponymist, the disease would be Landry–Guillain–Barré–Strohl syndrome. The precise and best name for the diagnosis in the majority of GBS patients is AIDP, acute inflammatory demyelinating polyneuropathy, which describes the disease well and allows more precise classification, but eponyms in neurology die hard.

2 Etiopathogenesis

In one of Landry's initial cases, his senior physician diagnosed one of his patients as having hysteria; time would prove this to be clearly incorrect. Another theory by Landry's contemporaries was that it was a variant of diphtheria. This diagnosis was closer to being correct since both diphtheritic neuropathy and GBS are demyelinating neuropathies (see "demyelination" in Chap. 2). Neurologists ultimately found them to be distinct diseases when the causative pathogen of diphtheria was found to be *Corynebacterium diphtheriae* by Friedrich Loeffler in 1884 (see Inset 9.2) , and patients with GBS were found to not be infected. Subsequently, GBS has been thought to be immune-mediated, due to a response against a "stimulating agent", usually an infection; this hypothesis is based on the epidemiology, i.e., its occurrence frequently after bacterial or viral infections, and on investigations in experimental autoimmune neuritis (EAN), GBS's animal model.

A.R. Pachner, *A Primer of Neuroimmunological Disease*,
DOI 10.1007/978-1-4614-2188-7_9, © Springer Science+Business Media, LLC 2012

Inset 9.2 Diphtheria and Its Recent Resurgence

Within the differential diagnosis of GBS is diphtheritic neuropathy caused by infection with the bacteria *Corynebacteria diphtheriae*. Brain's textbook *Diseases of the Nervous System* in its seventh edition published in 1969, devoted three pages to diphtheritic neuropathy, and an equal amount to GBS. However, most American neurologists, though they will diagnose many cases of GBS in their lifetimes, will never see diphtheritic neuropathy in their practices because diphtheria vaccination, usually in combination with pertussis and tetanus, i.e. the DPT vaccine, is required for school-age children in the USA, and there is almost no diphtheria in the USA. There were only three cases of diphtheria in the USA between 2000 and 2007, an amazing contrast to the 1920s where there were estimated to be more than 100,000 cases per year in the USA. Thus, the absence of diphtheritic neuropathy represents a triumph of public health vigilance; diphtheria will return if this vigilance fails. Such an event occurred in the former Soviet Republics after the breakup of the Soviet Union when diphtheria vaccination rates dropped. By 1998, there were as many as 200,000 cases of diphtheria in the former Soviet Union countries with 5,000 deaths. This resurgence of diphtheria also led to a predictable increase in cases of diphtheritic neuropathy (DN), which was analyzed in depth by Logina and Donaghy [12] who compared clinical features of 50 adults with DN and contrasted them to 21 patients with GBS. DN is thought to be due to demyelinative effects of the diphtheria toxin. The first Nobel Prize in Medicine in 1901 was awarded to Emil von Behring for his development of diphtheria antitoxin, which is still used and is highly effective in preventing diphtheritic neuropathy if given early in the course of the infection.

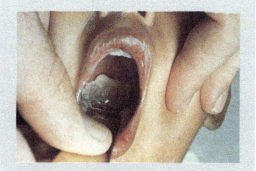

Fig. 9.1 Diphtheritic membrane. Nasopharyngeal diphtheria with extensive membrane formation in a 4-year-old child

DN usually presents with palatal paralysis as the first manifestation while symptoms of generalized polyneuropathy do not develop until usually weeks later; rarely, generalized weakness can be the first manifestation of DN. There is commonly the appearance on examination of the mouth and throat, a diphtheritic "membrane" (Fig. 9.1). Problems with dysarthria are common in DN, because of early involvement of the palate and uvula. Otherwise, DN and GBS can be very similar, both demonstrating severe hyporeflexive flaccid weakness. However, because diphtheria toxin can have its effect locally early, paralysis of bulbar muscles, i.e. muscles innervated by the cranial nerves, is very common in DN, occurring in 98% of DN patients in Logina's series, while it is unusual in GBS, occurring in only 10% of her GBS cases. Even DPT-vaccinated individuals can have an attenuated form of DN, if they are exposed to the bacteria; thus, any individual who has a "bulbar form of GBS" should be considered to possibly have DN if they have traveled to countries of the former Soviet Union or other areas where diphtheria is endemic.

The disease is not unique to humans and occurs naturally in dogs (e.g., coonhound paralysis) and in nonhuman primates. Coonhound paralysis appears to be due to an immune reaction to proteins present in the saliva of raccoons and requires previous exposure to raccoons in susceptible dogs; the disease does not occur in cats. It occurs 1–2 weeks after exposure to the raccoon, similar to the latency in humans after exposure to the "stimulating agent." Affected dogs initially have gait problems progressing to weakness of all limbs, usually beginning in the hind limbs then moving forward to the forelimbs. Deep tendon reflexes (DTRs) are lost and respiratory paralysis may develop.

3 Pathology

There are few studies on pathology in GBS because most patients with GBS do not undergo nerve biopsy. When tissue is obtained, pathology on gross examination is not usually abnormal, but microscopic analysis can show demyelination and frequently inflammation in nerve roots and peripheral nerves. The mononuclear cell infiltrate was one of the first hints that the underlying etiology was likely immune-mediated. Demyelination is found predominantly near nodes of Ranvier (see Chap. 2 for more on nodes of Ranvier) often accompanied by the presence of inflammatory cells. Some forms of GBS are axonal, and axonal degeneration without demyelination can be seen on pathological evaluation in these varieties. Occasionally, myelitis, i.e. inflammation of the spinal cord, can be an associated finding.

4 Epidemiology and Etiopathogenesis

There is no known genetic predisposition to GBS. There may be an increased incidence with increasing age. Most cases of GBS have an antecedent infection within the previous few weeks, and there has been increasing evidence that a triggering pathogen is the bacterium *Campylobacter jejuni*, which is a common pathogen causing infectious diarrhea. In China, GBS, particularly the AMAN form (see below), is epidemic, occurring in the summer months, and is often associated with an antecedent diarrheal infection, commonly due to *C. jejuni* [1]. Initially, it was hypothesized that a phenomenon termed molecular mimicry was responsible for the GBS. That is, after the human developed *C. jejuni* infection, and developed antibodies to glycolipids on the surface of the *C. jejuni* bacteria, including the glycolipid GM1, these antibodies caused an immune attack on the GM1 in the GBS patient's nerve tissue. However, this initial hypothesis could not be definitively proven and considerable evidence against it has emerged.

However, the hypothesis that GBS is due to pathogenic molecules in the serum of GBS patients has been confirmed to some extent by the beneficial effect of plasmapheresis, a procedure in which large volumes of the patient's plasma are replaced with normal saline (see treatment, below). Many investigators feel that the pathogenic material is the immunoglobulin fraction, specifically antibodies to glycolipids, i.e., nonprotein molecules that are combinations of carbohydrate and lipid moieties. A large group of molecules called gangliosides, which are highly enriched in myelin, have undergone extensive study as being important in the pathogenesis of GBS. These investigations have not yielded a great deal of clinical benefit as yet, although one type of GBS, the Miller Fisher variant (see classification below), is reported as having positive anti-GQ1b antibodies, a type of anti-ganglioside antibody in 85% of affected patients. Many researchers in this field feel that anti-GQ1b antibodies are indeed pathogenic and there is some evidence for this in the research laboratory.

GBS usually attacks myelin without directly damaging neurons. Remyelination is likely an early feature of GBS, but is initially overwhelmed by damage to myelin. Recovery can be relatively rapid once continued damage to myelin halts and remyelination continues. Most patients with GBS ultimately recover completely even without treatment. The axon remains healthy in most patients with GBS, and simple remyelination by Schwann cells is the only requirement, a process which is usually relatively rapid, i.e., weeks to a few months. This rapid recovery is, in contrast, to processes which damage the axon, in which case Wallerian degeneration (see Chap. 2) can occur, and complete recovery frequently does not occur.

had a severe case requiring one and half months in an intensive care unit. The incidence of the disease is low, only about 2 cases per 100,000 population per year, but the disease is extremely important since it can be fatal if not diagnosed early and treated by respiratory support.

5   Clinical Manifestations

An excellent description of the disease is by Landry who first described GBS in 1859:

> the main problem is usually a motor disorder characterised by a gradual diminution of muscular strength with flaccid limbs and without contractures, convulsions or reflex movements of any kind. In almost all cases micturition and defaecation remain normal. One does not observe any symptoms referable to the central nervous system, spinal pain or tenderness, headache or delirium. The intellectual faculties are preserved until the end. The onset of the paralysis can be preceded by a general feeling of weakness, pins and needles and even slight cramps. Alternatively the illness may begin suddenly and end unexpectedly. In both cases the weakness spreads rapidly from the lower to the upper parts of the body with a universal tendency to become generalised.
>   The first symptoms always affect the extremities of the limbs and the lower limbs particularly… The paralysis then becomes generalised but more severe in the distal parts of the extremities. The progression can be more or less rapid. It was 8 days in 1 and 15 days in another case which I believe can be classified as acute. More often it is scarcely 2 or 3 days and sometimes only a few hours.
>   When the paralysis reaches its maximum intensity the danger of asphyxia is always imminent. However in eight out of ten cases death was avoided either by skillful professional intervention or a spontaneous remission of this phase of the illness. In two cases death occurred at this stage . . . When the paralysis recedes it demonstrates the reverse of the phenomenon which signaled its development. The upper parts of the body, the last to be affected, are the first to recover their mobility which then returns from above downwards." (from the Guillain–Barré Syndrome Support Group of UK and Ireland's web site: http://www.gbs.org.uk/history.html).

Another excellent description of the disease from the patient's viewpoint is the book, *No Laughing Matter*, written by Joseph Heller and his friend Speed Vogel, after Heller contracted GBS in 1981. Heller, best known as the author of *Catch-22*,

6   Natural History and Prognosis

Most patients with GBS, even those with severe forms requiring intubation to maintain respiration, have full or nearly full recoveries, which are accelerated by plasmapheresis therapy. For instance, those who presented to the hospital with weakness so severe that they required respiratory support required 6 months of recovery before they could walk, while those who did not require respiratory support during the worst of their GBS required only 2 months of recovery. Treatment with plasmapheresis improved those numbers to 3 and 1.5 months, respectively [2]. Death from respiratory failure, relatively common in Landry's time, is now unusual because of the availability of ventilators.

7   Classification

The most common form occurring in about 95% of patients with GBS in the USA and Europe is AIDP, the "classical" form which is primarily demyelinating; in this form axonal function is spared, and the prominent demyelination is responsible for most of the neurological deficit. In 1991, McKhann and colleagues described a new form of GBS which came to be called acute motor axonal neuropathy (AMAN) [3]. AMAN was distinct from AIDP in being primarily seen in epidemics in children in rural China and involving axonal function on electrophysiological testing (Table 9.1). In neither AIDP nor AMAN is sensory involvement prominent, while in another form of GBS, called acute motor and sensory axonal neuropathy (AMSAN), sensory involvement is more obvious. Still another form is the Fisher variant of GBS, sometimes called the Miller Fisher syndrome or MFS, described by the neurologist C. Miller Fisher in 1956, in which the triad of ophthalmoplegia,

Table 9.1 Characteristics of AIDP and AMAN

Characteristic	AIDP	AMAN
Found in	Adults	Children
Geography	World	Rural China
Occurrence	Sporadic	Epidemic
Electrophysiology	Demyelinating	Axonal

ataxia, and areflexia is present with limb weakness being absent or less prominent. Overlap syndromes of the various forms of GBS are not rare, with varying motor vs. sensory involvement, limb weakness, respiratory involvement, ophthalmoplegia (inability to move the eyes), facial and other cranial nerve palsies, and sensory neuropathies. AIDP is generally not associated with any characteristic autoantibody, while the other forms of GBS have been found more commonly associated with anti-ganglioside antibodies. In the majority of patients with MFS, anti-GQ1b antibodies are present. In this respect, MFS resembles another inflammatory syndrome called Bickerstaff's brainstem encephalitis. In the 1950s, Bickerstaff reported eight patients with impaired consciousness of which all had ataxia, seven had ophthalmoplegia, and four hypo- or areflexia. From the time of their first descriptions, MFS and Bickerstaff's were considered to be related, and the later finding that anti-GQ1b antibodies were present in high percentages of both diseases supported the clinical resemblance.

8 Diagnosis

The diagnosis of AIDP, the most common form of GBS, is usually straightforward for neurologists. Symptoms consist of motor weakness out of proportion to sensory symptoms with an acute onset and rapid progression usually over a few days, but rarely over hours or over a few weeks. The weakness is frequently ascending from the legs to the arms to bulbar muscles. Respiratory involvement is common, frequently requiring intubation for respiratory support. On examination, there is motor weakness with absent or minimal sensory findings. DTRs are usually absent, rarely can be present, but should be markedly diminished. Cerebrospinal fluid analysis reveals

the "dissociation albumino-cytologique" first described by Guillain, Barré, and Strohl in 1916, i.e., markedly elevated protein with normal or only slightly elevated white cell count. Electromyography and nerve conduction studies reveal prolonged nerve conductions, proximally and distally, consistent with demyelination. Very few diseases mimic AIDP, but other causes of acute neuropathy should be considered such as toxic neuropathies, Lyme disease, or vasculitis; rare conditions such as diphtheria, tick paralysis, or porphyria are in the differential.

9 Therapy

The therapy over the last century which has made the biggest impact on improving care for GBS patients was the development of ventilators, beginning with the negative-pressure "iron lungs" in the early twentieth century to the modern positive pressure devices, to support ventilation impaired by the weakness of the disease. Ventilators in addition to modern advances in care of critically ill patients in intensive care units has saved untold lives in patients with severe GBS. Since the natural history of GBS is usually one of complete recovery, these critically ill patients, after their harrowing experience with the disease, usually return to normal lives.

Two immunoactive treatments, plasma exchange and intravenous immunoglobulin G (IVIg), have been shown to be effective in shortening the duration of weakness, while another, corticosteroids, has been shown to be ineffective (see Chap. 19 for a description of these treatments). Because of a large Dutch study comparing plasma exchange and IVIg which demonstrated at a minimum, the equivalence, and possibly superiority of IVIg over plasma exchange [4], many clinicians use IVIg at the dose used in the Dutch study, 0.4 g/kg/day over 5 days. The use of IVIg in the treatment of GBS has been recently reviewed in depth for the Cochrane reviews [5]. Given the efficacy of Eculizumab, an anti-complement monoclonal antibody, in a mouse model of GBS [6], agents targeting complement appear ready for clinical studies.

10   GBS's Cousin: Chronic Inflammatory Demyelinating Polyneuropathy

Inflammation within nerves associated with injury can be constant not just transient. GBS has a cousin, chronic inflammatory demyelinating polyneuropathy (CIDP), in which inflammation is persistent, or possibly recurrent, and the injury is not transient as it is in AIDP. Criteria for the diagnosis of CIDP are not universally accepted, and there have been many different iterations through the years. A recent consensus conference defined CIDP as

> a chronic non-genetic polyneuropathy, progressive for at least eight weeks, without a serum paraprotein and either 1) recordable compound muscle action potentials in > or =75% of motor nerves and either abnormal distal latency in >50% of nerves or abnormal motor conduction velocity in >50% of nerves or abnormal F wave latency in >50% of nerves; or 2) symmetrical onset of motor symptoms, symmetrical weakness of four limbs, and proximal weakness in > or =1 limb. [7].

This definition excludes AIDP because of the stipulation of being chronically progressive over at least 8 weeks. Although this definition allows the diagnosis of CIDP to be made without electrophysiological analysis, most neurologists will want electromyography/nerve conduction velocity (EMG/NCV) confirmation, since the diagnosis of CIDP generally commits the patient and neurologist to prolonged therapy.

CIDP is a relatively common disease, with a prevalence of approximately 5–10 cases per 100,000 population. In CIDP as in AIDP, plasma exchange and IVIg are effective [8]. Additionally in CIDP, in contrast to AIDP, corticosteroids are effective. In patients with CIDP, the challenge is the appropriate diagnosis, since many patients can be misdiagnosed and treatment is thus delayed. The natural history of CIDP is variable; some patients, the minority, respond to initial treatment and achieve complete remission, while the majority of patients relapse or progress and require prolonged treatment [9].

11   Neuropathy Associated with Monoclonal Gammopathy

Immunoglobulin, as described in Chap. 1, is present in blood normally as billions of different molecules, composed of millions of clonal populations. Occasionally, one clone can predominate, i.e., monoclonal gammopathy (see Inset 9.3), both in patients with diseases such as multiple myeloma, a B-cell malignancy, and in apparently healthy individuals; the latter of these two situations is sometimes called benign monoclonal gammopathy or more appropriately monoclonal gammopathy of unknown significance (MGUS). Neuropathies are a more frequent occurrence in patients with monoclonal gammopathies, either in MGUS or in myelomas, than in control groups, and are usually associated with gammopathies of IgM, rather than IgG or IgA. The incidence, etiopathogenesis, pathology, natural history, and optimal treatment are unknown. 1% of the population over the age of 25 has MGUS, most of them being IgG gammopathies.

Inset 9.3 Monoclonal Gammopathy

The words "monoclonal gammopathy" describe the condition: "monoclonal" means a single clone, and "gammopathy" indicates an abnormal condition of γ-globulins, a type of protein in serum that is large, larger than albumin, the α- and the β-globulins. The most important γ-globulins are the immunoglobulins, the antibody molecules produced by the highly specialized B lymphocytes called plasma cells. A monoclonal gammopathy occurs when a single plasma cell clone is overproducing one specific immunoglobulin.

Proteins in serum are measured by laboratories using a serum protein electrophoresis (SPEP), in which the proteins are separated by size and the information displayed on a curve in which the y-axis represents amount of protein and the x-axis represents size. The lowest-sized protein peak is albumin, containing the largest amount of protein, while the γ-globulins have a broad peak representing the diversity of billions of different molecules of various large size. In the abnormal SPEP from a patient with a monoclonal gammopathy, the albumin, α- and β-globulin peaks are normally shaped, but the γ-globulin region has a large uniform peak representing an M-protein, sometimes called M-gradient, a monoclonal protein. If the presence of the M-protein is not associated with any known disease, the finding is called monoclonal gammopathy of unknown significance or MGUS; frequently, an M-protein is found in cases of the plasma cell malignancy called multiple myeloma. In patients with the neuropathy associated with monoclonal gammopathy, it is believed that the M-protein has reactivity to a component of nerve and, by binding to it and activating downstream immune effectors such as macrophages, the M-protein results in nerve injury.

There are interesting subgroups within these disorders. For instance, a very high percentage of patients with the osteosclerotic form of multiple myeloma develop neuropathy. In addition, some of these patients develop a syndrome called POEMS, in which the P (peripheral neuropathy) and M (monoclonal gammopathy) are also associated with organomegaly (O), endocrinopathy (E), and skin changes (S). Most patients with neuropathy and MGUS have a sensory neuropathy called distal acquired demyelinating sensory neuropathy (DADS), which is very slowly progressive and associated with minimal weakness. In a large percentage of DADS patients with IgM gammopathies, the paraprotein reacts with the myelin protein, myelin-associated glycoprotein

(MAG). MAG (Inset 9.4) is a protein found in both the CNS and the PNS myelin and is a member of the family of proteins called the immunoglobulin superfamily (IgSF). The mechanism by which a high titer anti-MAG antibody might produce a neuropathy has not been worked out, but MAG is thought to be an important molecule for interactions between myelin and the axon.

Inset 9.4 MAG: A Member of the Immunoglobulin Superfamily

MAG is a 100 kDa protein in the myelin sheath near axons, mostly present in noncompacted myelin, and is thought to possibly function in interactions between glia and axons in both the CNS and the PNS. It is a member of the immunoglobulin superfamily, sometimes abbreviated IgSF, because the members of the family bear structural homologies to immunoglobulin, especially the presence of immunoglobulin domains resembling the Fab portion of immunoglobulins. Most members of the IgSF are associated with the immune response, such as cytokine receptors, adhesion molecules, and proteins involved in the major histocompatibility complex.

MAG on glial cells is an inhibitor of neurite outgrowth presumably through binding to a variety of receptors on neural cells such as p75, a nerve growth factor receptor, or carbohydrates on the neural surface. The IgM from patients with monoclonal gammopathy and neuropathy which binds to MAG also binds to other molecules that share carbohydrate epitopes with MAG such as P0, PMP-22, and glycolipids such as sulfate-3-glucuronly paragloboside (SGPG). At this time, it is unknown which of the above target molecules may be critical for the neuropathogenicity of the M-protein in patients with neuropathy and monoclonal gammopathy.

12   Experimental Neuritis

EAN was first described by Byron Waksman and Raymond Adams in 1955 when they immunized rabbits with homogenized peripheral nerves and found the animals developed neuritis. EAN can be induced in a variety of animals by immunization with any number of peripheral nerve components, including whole peripheral nerve myelin or the peripheral myelin proteins P_0, P_2, and PMP_{22}. Susceptible animals have been the mouse, guinea pig, and rabbit, but the most commonly utilized animal is the rat, usually the Lewis rat. Within 7–14 days after immunization, animals become weak, presumably from segmental demyelination. This damage is thought to be due to anti-myelin T-cell responses, mostly in the nerve root, associated with both infiltrating and endoneural resident macrophages. As in EAE, EAN is a monophasic event and animals recover after the acute event, similar to GBS. Some models can be transferred with "neuritogenic" T cells in a manner similar to the transfer of EAE with "encephalitogenic" T cells, and it is thought that Lewis rat EAN is primarily T-cell-mediated. Cells and large molecules infiltrating into nerve must pass through a barrier called the blood–nerve barrier (BNB), which is similar conceptually to the blood–brain barrier. Demyelinating forms of EAN can also be caused by "bystander demyelination," e.g., an immune response to ovalbumin, an irrelevant antigen, within the nerve; i.e., in an ovalbumin-sensitized mouse, injection of ovalbumin into the nerve can induce a local experimental neuritis. Cell adhesion molecules, chemokines, matrix metalloproteinases, cytokines, macrophages, antimyelin antibodies, complement, and possibly Schwann cells are also all thought to contribute to full expression of EAN. After monophasic weakness, recovery follows usually occurring over a few weeks, possibly related to apoptosis, programmed cell death, of infiltrating T cells.

The precise molecules targeted by the immune attack in various immune-mediated neuropathies has been a subject of intense study. Proteins of peripheral nerve myelin such as P_0, P_2, and PMP_{22} have been the standard immunogens but glycolipids, particularly gangliosides, have also been used. Gangliosides are complex glycolipids, nonprotein compounds which contain carbohydrates linked to lipid molecules, with negatively charged oligosacchrides and one or more sialic acid residues. Many different models of autoimmune neuritis have been described based on immunity to gangliosides, but none has proven sufficiently reproducible and robust to be used extensively. A particularly intriguing ganglioside immunogen has been sulfated glucuronyl paragloboside (SGPG), which has been used as an immunogen to induce neuropathy in rabbits, rats, and cats [10, 11]. The most convincing ganglioside antigen linking experimental to human disease is G1Qb. Antibodies to this ganglioside have been found in a large percentage of patients with the Miller Fisher variant of GBS, and Halstead et al. [6] have described in vitro and in vivo models of neuropathy induced by a monoclonal anti-G1Qb antibody. Models of anti-ganglioside immune-mediated appear to be dependent on anti-ganglioside antibody, often IgM, and complement, rather than being T cell mediated. Gangliosides other than SGPG and G1Qb can be targets of immune attacks, but the participation of any particular ganglioside in immune-mediated neuropathies appears to be at least partly due to their accessibility in vivo to circulating pathogenic antibodies; many ganglioside are cryptic antigens, i.e., not expressed on the surface of nerve.

There are fewer investigators working on EAN than EAE for at least two reasons. First, inflammatory neuropathies in humans are less common and less debilitating generally than is multiple sclerosis, so the pressure for translational research is less. Second, EAN and ganglioside-mediated neuropathies are most easily induced in rats or rabbits, not mice, and thus the wealth of mutant mice available for research in EAE is not as easily accessible to EAN researchers. The latter problem has been recently attempted to be addressed by a few researchers who have induced experimental neuropathies in mice.

References

1. Ho TW, Mishu B, Li CY, et al. Guillain-Barre syndrome in northern China. Relationship to Campylobacter jejuni infection and anti-glycolipid antibodies. Brain. 1995;118(Pt 3):597–605.
2. McKhann GM, Griffin JW, Cornblath DR, Mellits ED, Fisher RS, Quaskey SA. Plasmapheresis and Guillain-Barre syndrome: analysis of prognostic factors and the effect of plasmapheresis. Ann Neurol. Apr 1988;23(4):347–53.
3. McKhann GM, Cornblath DR, Ho T, et al. Clinical and electrophysiological aspects of acute paralytic disease of children and young adults in northern China. Lancet. 1991;338(8767):593–7.
4. van der Meche FG, Schmitz PI, Dutch Guillain-Barre Study Group. A randomized trial comparing intravenous immune globulin and plasma exchange in Guillain-Barre syndrome. N Engl J Med. 1992; 326(17):1123–9.
5. Hughes RA, Raphael JC, Swan AV, van Doorn PA. Intravenous immunoglobulin for Guillain-Barre syndrome. *Cochrane Database Syst Rev.* 2006(1): CD002063.
6. Halstead SK, Zitman FM, Humphreys PD, et al. Eculizumab prevents anti-ganglioside antibody-mediated neuropathy in a murine model. Brain. May 2008;131(Pt 5):1197–208.
7. Koski CL, Baumgarten M, Magder LS, et al. Derivation and validation of diagnostic criteria for chronic inflammatory demyelinating polyneuropathy. J Neurol Sci. 2009;277(1–2):1–8.
8. Eftimov F, Winer JB, Vermeulen M, de Haan R, van Schaik IN. Intravenous immunoglobulin for chronic inflammatory demyelinating polyradiculoneuropathy. *Cochrane Database Syst Rev.* 2009(1):CD001797.
9. Barohn RJ, Kissel JT, Warmolts JR, Mendell JR. Chronic inflammatory demyelinating polyradiculoneuropathy. Clinical characteristics, course, and recommendations for diagnostic criteria. Arch Neurol. 1989;46(8):878–84.
10. Yu RK, Usuki S, Ariga T. Ganglioside molecular mimicry and its pathological roles in Guillain-Barre syndrome and related diseases. Infect Immun. Dec 2006;74(12):6517–27.
11. Ilyas AA, Gu Y, Dalakas MC, Quarles RH, Bhatt S. Induction of experimental ataxic sensory neuronopathy in cats by immunization with purified SGPG. J Neuroimmunol. Jan 2008;193(1–2):87–93.
12. Logina I, Donaghy M. Diphtheritic polyneuropathy: a clinical study and comparison with Guillain-Barre syndrome. J Neurol Neurosurg Psychiatry. Oct 1999;67(4):433–8.

1 Definition

Myasthenia gravis (derived from "my"-muscle, "asthenia"-weakness, and "gravis"-severe) is a neuromuscular disease whose clinical hallmark is fatigable muscle weakness caused by autoantibodies to proteins at the neuromuscular junction (NMJ).

2 Etiopathogenesis

The NMJ is a highly specialized anatomical intersection between motor nerves and muscle and contains the motor nerve terminal, a space between the terminal and the muscle, and a complex section of the muscle membrane containing a number of proteins not present at other parts of the muscle. The pathogenesis of most forms of myasthenia gravis (MG) is an antibody-mediated attack on the acetylcholine receptor (AChR). These antibodies cause impaired function of the AChR and thus the NMJ, leading to fatigable weakness. It is likely that at least three different mechanisms account for the damage. One is complement-dependent lysis of AChRs, another is cross-linking of receptors and subsequent degradation, and the third is direct blockade of the ion channel. Most patients with severe MG have a combination of all three mechanisms, although most investigators feel that complement-dependent lysis leading to loss of AChR numbers is the predominant mechanism responsible for the clinical symptoms. This ultimately results in a loss of AChRs and a pathological simplification of the complex frond-like architecture of the postsynaptic muscle membrane as shown in Fig. 10.1.

The particular type of AChR targeted by antibodies in the serum of MG patients is the skeletal muscle nicotinic AChR. This form of the AChR is distinct from AChRs present in the central nervous system, which consist of both nicotinic and muscarinic receptors with different structures. The nicotinic AChR is an oligomeric membrane protein, found in the normal adult almost exclusively at specialized areas on muscle membranes called NMJs, which are adjacent to axonal terminals. The protein consists of five subunits, two alpha, and one each of beta, gamma, and epsilon subunits. It functions as a ligand-gated ion channel. The release of adequate acetylcholine (ACh) at the axon terminals, diffusion across the NMJ, and binding to the AChR in the postsynaptic membrane results in an action potential in the muscle; this is impaired in MG by the anti-AChR antibodies. The impaired function of AChRs at the end plate is responsible for the symptoms in MG patients.

Under normal conditions there is a balance between the continued release of ACh from the presynaptic terminal and destruction of ACh within the NMJ by an enzyme called acetylcholinesterase. Acetylcholinesterase is a critical molecule for neuromuscular transmission since ACh, once released into the NMJ, is destroyed quickly if it does not bind to the AChR.

A.R. Pachner, *A Primer of Neuroimmunological Disease*,
DOI 10.1007/978-1-4614-2188-7_10, © Springer Science+Business Media, LLC 2012

Fig. 10.1 Three mechanisms by which autoantibodies to a receptor cause loss of the channel from the surface. Loss of AChRs and change in muscle membrane in myasthenia gravis

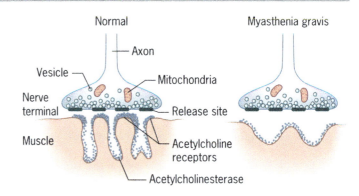

Increased ACh at the NMJ, achieved by inhibiting acetylcholinesterase, sometimes will improve NMJ function to make up for lost AChRs, which is the basis for acetylcholinesterase inhibitors used in the diagnosis and treatment of MG (see below). Conversely, too much ACh is harmful since it can lead to desensitization of the AChR and weakness or paralysis.

3   Pathology

Neurologists and neuropathologists of the late nineteenth and early twentieth centuries were mystified by the disparity between the profound clinical symptoms and signs in MG and the absence of pathological findings that would explain them. Many patients did die of the disease, hence the name gravis, from the Latin for "weighty, serious, or heavy," so there was no shortage of autopsy material. Occasionally, pathologists would see collections of lymphocytes in muscle, called lymphorrhages, but these were neither sensitive nor specific for the disease. It was not until the link to the AChR was clear, and pathologists could use specific probes for the AChR, that it was clear that myasthenics had markedly decreased AChR at the NMJ.

The specific probe used in labeling the AChR is bungarotoxin, a toxin produced by the Indonesian krait *Bungarus multicinctus*, which binds nearly irreversibly to the AChR. This toxin

is used not only for identifying the pathology of MG in muscle biopsies but also in the immunoassay for the characteristic autoantibody (see Sect. 8). Other snake toxins such as cobratoxin are also used in AChR and MG research (see Inset 10.1, Animal toxins in neurological research).

Inset 10.1 Animal Toxins in Neurological Disease, Including Conotoxin

Much of the publicity for animal toxins in neurological disease has been on toxins from snakes such as the Indonesian many-banded krait (bungarotoxin from *Bungarus multicinctus*) and cobra (cobratoxin from *Naja naja*) which bind to the nicotinic AChR, or the mamba (dendrotoxin from *Dendroaspis*) which binds to voltage-gated potassium channels. However, toxins from snails, spiders, and wasps are up-and-comers. Omega-conotoxin is the first non-snake toxin to be used extensively in a clinical setting, as a probe for VGCCs to detect the anti-VGCC antibodies in LEMS. Cone snails prey on other animals including fish and use their toxins to immobilize their prey. They use a specialized form of tooth, called a toxoglossan radula, as a barbed harpoon which has the conotoxins in channels. Omega-conotoxin is not only used in the anti-VGCC antibody assay but also is the model for the chemically produced drug, Ziconotide, which is FDA approved for intrathecal use for the treatment of pain. Spiders have a number of toxins. Atracotoxins operate by opening sodium channels while agatoxins from the funnel web spider (*Agelenopsis aperta*) have a number of effects on various ion channels.

4 Genetics and Epidemiology

There are no known genetic proclivities for this disease, except that the disease is associated with other autoimmune diseases and frequently runs in families in which other members have other autoimmune conditions such as systemic lupus erythematosus, thyroiditis, rheumatoid arthritis, or ankylosing spondylitis.

When the disease begins in young individuals, it mostly occurs in females, and is usually associated with thymic hyperplasia, a condition in which the thymus (see Chap. 1) is enlarged. The link between thymic hyperplasia and MG remains unexplained. MG is increasingly being diagnosed in people over 50, in patients with a high incidence of thymic neoplasms, mostly benign, called thymomas. Patients with thymomas frequently have autoantibodies to other muscle proteins, such as titin and the ryanodine receptor. There may be some geographic peculiarities to MG epidemiology. Zhang et al. [1] reported a high incidence of childhood MG in China's Hubei province.

5 Clinical Manifestations

The hallmark of MG is fatigable muscle weakness (see Inset 10.2, a 28-year-old woman with fatigue, misdiagnosed as having Lyme disease). Sustained effort in an involved muscle results in decreased ability of the muscle to contract, but after rest the muscle might be able to function normally for a short time before it begins to fatigue again. Thus, a myasthenic patient's

Inset 10.2 28-Year-Old Woman with Fatigue, Misdiagnosed as Having Lyme Disease

This 28-year-old mother of two had been to many doctors for severe fatigue, usually at night, and occasional problems with speech and swallowing. She came to UMDNJ because of our expertise in Lyme disease to confirm the diagnosis of Lyme disease which had been made by a community physician. She had no clinical manifestations of Lyme disease, but one of many blood tests had been borderline positive. She had undergone extensive intravenous antibiotic therapy for Lyme disease without clear benefit. As she told her story, her speech became more and more slurred. After a few minutes of rest, her speech was normal, but then slurred again as she spoke more. The diagnosis of MG was entertained, then confirmed, and treatment resulted in complete resolution of her symptoms. She has been symptom-free, except for occasional mild flares of trouble with speech, for 10 years with treatment mostly with cholinesterase inhibitors.

examination might change a number of times through the day, depending on the prior use of involved muscles. Some muscles are commonly involved, such as the extraocular muscles (EOMs), facial muscles, muscles of speech and swallowing, or proximal muscles of the arm or leg, while distal muscles are less frequently involved. Thus, common symptoms are double vision, drooping eyelids, slurred speech, difficulty combing hair, or walking upstairs. All of these symptoms worsen with repetition. Thus, a patient may start out speaking a few sentences normally, but as the speech extends to more than a few seconds, slurring begins. Patients tend to be more symptomatic at night, as their muscles fatigue during the day, and they can appear relatively normal in the morning. Since patients frequently see physicians during the morning hours, MG patients will frequently be initially diagnosed as having depression, hysteria, or malingering.

Most of the time the symptoms are chronic or subacute, fluctuating, and unpredictable. Occasionally, usually with an infection or associated with surgery, patients experience a myasthenic crisis, in which the symptoms exacerbate rapidly and deterioration can be severe and life-threatening. During these crises, maintenance of swallowing and respiration is critical, and patients usually need to be monitored in an intensive care unit.

6   Natural History and Prognosis

The natural history of MG is highly variable. Some patients, especially those with the ocular form, are inconvenienced by the disease, but not significantly disabled. In contrast, patients with severe generalized MG, especially if the weakness involves muscles of swallowing and breathing, can have an unremitting course over months to a few years with severe disability and ultimately death from respiratory weakness, infections, or aspiration pneumonia. It is generally believed that the first 5 years of the disease tend to be the most dangerous and if the neurologist can get the patient through that period, the remainder of the course is not progressive. This is in dramatic contrast to multiple sclerosis (see Chap. 4), the most common neuroimmunological disease, in which the disease course is progressive, and most disability accrues after the patient has had the disease for 5 years.

7   Classification

The disease has been divided into "ocular" myasthenia and generalized myasthenia. Ocular myasthenics have symptoms limited to their eye movements, with ptosis (eyelid drooping) and diplopia (double vision). The distinction is not complete, since most generalized patients have ocular symptoms and signs, and most ocular myasthenics have subclinical involvement of many other muscles that can be detected when

sensitive electromyographic testing is performed. Another method of classifying is by age: early-onset MG is primarily seen in females and is associated with thymic hyperplasia, late-onset MG being seen equally in both sexes and associated with a high incidence of thymomas.

Increasingly, neurologists are also classifying the patients on the basis of the autoantibody population involved. Those who do not have detectable antibodies to AChR by the standard immunoprecipitation assay are called "seronegative", but frequently have autoantibodies, detectable in serum, to proteins at the NMJ other than AChR. The most frequent autoantigen targeted by antibodies in seronegative MG is muscle-specific kinase (MuSK). Anti-MuSK antibody-positive patients have a different MG phenotype, with more facial involvement, both clinically and by electromyographic testing. Sometimes facial muscle atrophy is evident. In addition, the thymus tends not to be enlarged in MuSK antibody-positive patients. Some seronegative patients may actually be seropositive but require more sensitive assays available in the research lab to detect anti-AChR antibodies [2].

8   Diagnosis

The diagnosis of MG is made by history, examination, and laboratory testing. The most characteristic symptom is excessive fatigability, in which muscle strength initially is normal, but fades very quickly with use (see Inset 10.2). This is a helpful symptom for the diagnosis because it is relatively specific. In most other neurological diseases affecting motor strength, weakness is present from the onset of muscle use. Most patients with MG have ocular symptoms. The examination can be remarkably normal when the well-rested patient is seen in the morning; in this feature of profound fatigability, the disease is relatively distinct from other diseases affecting the nervous system. There is no significant sensory dysfunction in MG, and deep tendon reflexes are usually normal, or even brisk. Thus, many features of the exam are normal.

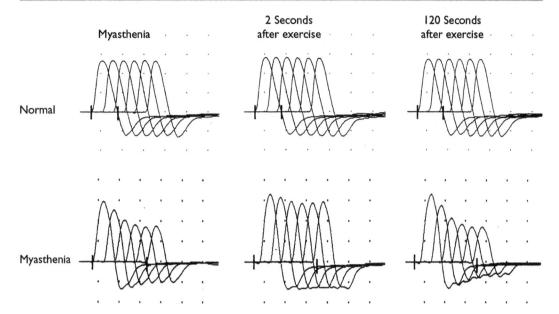

Fig. 10.2 Repetitive stimulation of muscle in normal individuals (*upper level*) and those with myasthenia gravis (*lower level*)

Objective findings of fatigable weakness can sometimes be elicited on examination. Many patients with MG have difficulty with sustained upgaze or tonic deviation of the eyes and develop ptosis (lid drooping) or an extraocular movement palsy when these maneuvers are performed. However, this takes patience by both the examiner and patient, since the weakness sometimes takes a minute or two to be manifested. Since hot temperatures exacerbate and cold temperatures ameliorate the weakness, an ice cube applied to a weak levator palpebrae muscle will sometimes reverse ptosis. Another test that has historically been done at the bedside for MG, the Tensilon test, is no longer feasible because of lack of commercial availability of Tensilon. In this test, edrophonium, a short-acting acetylcholinesterase inhibitor, was injected intravenously, resulting in a prompt temporary improvement of symptoms. The use of acetylcholinesterase inhibitors will be discussed in Sect. 9 below.

Most neurologists will utilize electromyography/nerve conduction (EMG/NCV) testing as an adjunct in the diagnosis of neuromuscular disorders, and MG is no exception. However, routine testing will not demonstrate abnormalities since both nerve conduction and routine muscle examinations are normal. When the diagnosis of MG is in the differential, the electromyographer will, in addition to routine tests, also perform repetitive stimulation to test transmission through the NMJ. Repetitive stimulation at 3–5 Hz, will demonstrate a "decremental response" (Fig. 10.2). At this rate of stimulation in normal individuals, the compound muscle action potential (CMAP) amplitude is the same for each stimulation, as shown on the left of the figure. In the myasthenic, in contrast, as shown on the right of the figure, the initial CMAP amplitude is normal, but subsequent potentials are lower and lower.

In generalized MG, most patients have detectable anti-AChR antibodies in their blood. The anti-AChR antibody assay is one of the few tests in neuroimmunology that has a nearly 100% specificity. Patients with a positive test have MG nearly 100% of the time and there are only very rare false positives. The test is said to be 85% sensitive in generalized MG and 50% sensitive in ocular MG, meaning that the test is false-negative 15 and 50% of the time, respectively. Other autoantibodies also frequently found in MG patients include anti-muscle antibodies, primarily to myofibrillar proteins such as actin, myosin or titin, antithyroid antibodies, and antibodies to voltage-gated potassium channels. These antibodies are not commonly used in clinical practice.

Given the above distinctive clinical features and the multitude of diagnostic aids, the competent neurologist will rarely err in diagnosing MG. In patients with early or mild disease when the symptoms are transient and the anti-AChR antibody negative, the diagnosis may be difficult. Some primary muscle diseases, such as unusual dystrophies or polymyositis (see Chap. 11), can be mistaken for MG, as can the Lambert–Eaton syndrome (see below). Some acute poisonings, such as with diphtheria or botulinum toxin, curare, snake bites especially cobras or kraits, can be confused with myasthenic crisis, not surprisingly since a number of these also target the AChR.

9　Therapy

Treatment for ocular myasthenia can be directed at the symptoms, such as patching one eye or surgery to keep eyelids raised, without incurring the risks and adverse effects of systemic therapy. Many myasthenics chose to utilize life-style changes, such as focusing activity early in the day and incorporating many rest periods. The safest form of pharmacological therapy is the use of anticholinesterases, the most popular of which is pyridostigmine. However, anticholinesterases are toxic if taken at too high a dose and the optimum dose varies from one patient to the next.

As with the Guillain–Barré syndrome (discussed in Chap. 9), the most dramatic improvement in the mortality of MG occurred with the widespread use of ventilators and intensive care units to provide respiratory support during the times of myasthenic crises. Because of these advances, MG is rarely a fatal disease today.

A range of immunotherapies has been utilized, including corticosteroids, azathioprine, cyclophosphamide, cyclosporin A, and mycophenolate, which are reviewed in Chap. 19. IVIg and plasmapheresis are also used, frequently just prior to surgery in MG patients to minimize the autoantibodies. None of these has been curative, but all are help-

ful in moderately to severely affected patients to improve symptoms. The risk/benefit ratio of each these agents must be carefully assessed, since none of these therapies is benign.

One of the most aggressive therapies utilized in MG is surgical removal of the thymus. The rationale for thymectomy is that the thymus is usually abnormal in MG, and removal will ameliorate the disease. Although this sounds logical, the evidence for benefit is not strong, and the potential risks of major thoracic surgery in patients with MG are significant. The complexities in consideration of this therapy in MG have been recently reviewed [3].

10　Lambert–Eaton Myasthenic Syndrome

In Lambert–Eaton myasthenic syndrome (LEMS), patients present with weakness but it is distributed differently than in MG, with trunk and leg involvement dominant, and ocular muscles only very rarely involved. In contrast to MG, symptoms are frequently worst in the morning upon awakening and improve with exertion. This pattern is also demonstrated by electromyography, in which there is initially a small action potential which increases with continued stimulation (see Fig. 10.3). The defect is the inability to release enough packets of ACh. Most patients with LEMS are middle-aged or elderly, in contrast to MG. Patients with LEMS usually have diminished deep tendon reflexes. About 50% of LEMS patients have small-cell lung cancer but other cancers are also associated. LEMS associated with cancers, like other paraneoplastic neurological syndromes (see Chap. 14) can present prior to the identification of the malignancy. All individuals who present with LEMS should be carefully evaluated for the presence of a cancer, since about 60% of patients with LEMS have an associated cancer.

In the majority of LEMS patients, there are autoantibodies to presynaptic voltage-gated calcium channels (VGCCs). These are detected by an immunoassay similar to that for MG, but a different toxin: omega-conotoxin from the marine

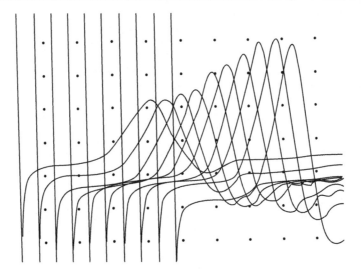

Fig. 10.3 Repetitive stimulation of muscle in a patient with LEMS

cone snail, which binds with very high affinity to VGCCs, is used to label the target protein (see Inset 10.1). Thus, the diagnosis of LEMS can usually be made on the basis of the characteristic signs, symptoms, and electromyography and confirmed with the demonstration of autoantibodies to the VGCCs. Many different types of treatment have been anecdotally tried in this rare condition, but the prognosis is determined by the underlying malignancy in the majority of LEMS patients.

11   Neonatal MG and Arthrogryposis Multiplex Congenita

Rarely, babies born to mothers with MG develop muscle weakness, presumably due to the placental transfer of maternal anti-AChR antibodies. Why this happens so infrequently in myasthenic mothers is unknown. Occasionally, babies with neonatal MG are born to women with preclinical MG, i.e., women who have anti-AChR antibodies but were asymptomatic. Some women with MG, especially those with high levels of antibodies to fetal AChR, deliver babies who are still-born or extremely weak, with fixed joints and other problems resulting from

decreased fetal movement, a condition called arthrogryposis multiplex congenita; arthrogryposis can also be caused by neonatal neuromuscular diseases other than MG. Improvements over the years in diagnosing and treating MG have made neonatal MG and arthrogryposis multiplex congenita less common.

12   Experimental Models of MG

EAMG (experimental autoimmune myasthenia gravis) is one of the animal models in neuroimmunological disease most faithful to the human condition, because the pathogenic entity in both the human and animal entity has been definitively demonstrated to be the same: a polyclonal population of anti-AChR antibodies. In EAMG, this population is usually elicited by immunization with a purified form of AChR, usually obtained from the specialized electric organ of electric eels such as the Pacific electric ray, Torpedo californica. The electric organ of these animals, which the animals use to stun and paralyze their prey, is highly enriched in AChRs. AChRs purified from electric rays were the first neurotransmitter receptors to be isolated and sequenced.

In the induction of EAMG, the electric organs of these animals is homogenized and then passed over a column in which the purified AChR-binding toxin from the king cobra *Naja naja* is bound. Once the AChR binds to the cobratoxin the nonbinding material is washed off, and the pure AChR can be eluted from the column. Experimental animals (mice, rats, or rabbits) can then be immunized with the eel AChR. After a time, often with boosting immunizations, the animal's immune response against the eel AChR begins to cross-react with the animal's own AChRs leading to the development of myasthenic weakness. Although weakness in the experimental animals is due to the action of anti-AChR antibodies at the NMJ, T cells specific for the AChR are required to help the B cells make antibody at adequate levels and with adequate affinity for the AChR. Using this approach in mice, monoclonal anti-AChR antibodies could be generated, which has aided in understanding AChR function.

A particularly important area in the AChR molecule is the main immunogenic region (MIR), an identical portion of each of the two alpha subunits in each AChR. A substantial portion of the anti-AChR antibodies in human and experimental MG bind to the MIR, which is a conformation-dependent region. The MIR is angled outward from the central axis of the AChR, and thus a single anti-AChR IgG antibody cannot crosslink the two subunits within a single molecule of AChR, but does crosslink adjacent AChRs. The MIR is distant from the ACh-binding site. Thus, anti-MIR antibodies do not block ACh binding to the NMJ. The site on the AChR which binds ACh is the same as that which binds snake toxins, and when antibodies bind there, they tend to be particularly pathogenic. Thus, it is likely that many of the anti-MIR antibodies in an anti-AChR antibody population do not markedly

interfere with AChR function, and thus it is not surprising that levels of anti-AChR antibodies do not predict disease severity. The target molecule is so large that an antibody could bind without necessarily affecting function.

In 1970s and 1980s, we saw a tremendous explosion of knowledge about the AChR and its role in EAMG. With this wealth of understanding, down to the molecular confirmation of the targets of the pathogenic antibodies, one might have predicted that it would not be too difficult to utilize this information toward therapy. Some promising efforts toward that end in EAMG have been recently summarized [4]. Unfortunately, new therapies based on this information have not been developed, and the current therapy remains nonspecific, broad spectrum immunosuppression of the whole immune response rather than a "smart bomb" targeted to a specific part of the AChR molecule. The reasons for this are complex, but a contributing factor is that MG is relatively a small market, and pharmaceutical companies may not be willing to make a large investment in research and clinical trials.

References

1. Zhang X, Yang M, Xu J, et al. Clinical and serological study of myasthenia gravis in HuBei Province, China. J Neurol Neurosurg Psychiatry. Apr 2007;78(4):386–90.
2. Brooks EB, Pachner AR, Drachman DB, Kantor FS. A sensitive rosetting assay for detection of acetylcholine receptor antibodies using BC3H-1 cells: positive results in 'antibody-negative' myasthenia gravis. J Neuroimmunol. 1990;28(1):83–93.
3. Richman DP, Agius MA. Treatment of autoimmune myasthenia gravis. Neurology. 2003;61(12):1652–61.
4. Pachner AR. Antigen-specific immunotherapy in myasthenia gravis: failed promise and new hope. J Neuroimmunol. 2004;152(1–2):vii–viii.

The Prototypic Neuroimmunological Muscle Disease: Dermatomyositis

11

1 Definition

Dermatomyositis is a rare inflammatory disease of muscle of unknown cause resulting in muscle weakness and pain in young- or middle-aged people. It is a member of a category of diseases that can be grouped together as idiopathic inflammatory myopathies (IIMs), the other members of which are polymyositis and inclusion body myositis (IBM).

2 Etiopathogenesis

The etiology is unknown but is assumed to be autoimmune because of the following combination of characteristics. First, there is no other explanation for the disease. Second, models resembling IIMs can be induced by immunization of animals with muscle proteins. Third, IIMs respond positively to anti-inflammatory medications. Fourth, there are a number of autoantibodies associated with dermatomyositis, which are divided into myositis-specific autoantibodies (MSAs), seen only in myositis, and myositis-associated autoantibodies, which can also be seen in other autoimmune diseases. However, none of the autoantibodies are found in the majority of patients, and the targets of these autoantibodies are all *intracellular* molecules, many of which are DNA-binding proteins. This is in contrast to myasthenia gravis where an autoimmune pathogenesis is well-accepted, mediated by autoantibodies to *extracellular* portions of the acetylcholine receptor. Thus, it is not clear whether the autoantibodies in IIMs are pathogenic, or simply secondary biomarkers of the inflammatory activity in muscle. Myositis can also be associated with malignancies as a paraneoplastic syndrome; this is discussed in Chap. 14.

Some recent data implicates a population of dendritic cells called plasmacytoid dendritic cells (PDCs) as being important in the IIMs (see Chap. 1 for a description of dendritic cells). These cells, when activated, produce large amounts of type I interferons, i.e., IFN-α and IFN-β. Biopsy specimens in patients with IIMs express large levels of interferon-induced genes such as the human myxovirus resistance 1 protein (MxA). The precise cause of the activation of PDCs in IIMs is unknown.

3 Pathology

Muscle biopsy reveals a mononuclear cell infiltrate which is usually in the perimysium, the thicker fibrous tissue separating muscle bundles. Degenerated muscle fibers may be present. There is atrophy of muscle fibrils at the edges of the muscle fascicle, so-called perifascicular atrophy. In the related myositis, polymyositis, there is an inflammatory exudate present in a different part of the muscle fiber, the endomysium, the thin layer of connective tissue around individual muscle fibers. In IIMs, there can be invasion of nonnecrotic muscle fibers by mononuclear cells, or other areas of inflammation such as around blood vessels.

A.R. Pachner, *A Primer of Neuroimmunological Disease*,
DOI 10.1007/978-1-4614-2188-7_11, © Springer Science+Business Media, LLC 2012

4   Genetics and Epidemiology

There are many genetic causes of myopathy, i.e., muscle disease, but the idiopathic *inflammatory* myopathies (IIMs), i.e., dermatomyositis, polymyositis, and inclusion-body myositis, do not appear to have a genetic component. These are rare diseases, with an annual incidence of approximately 1 in 100,000. Most studies find that more women than men have dermatomyositis and polymyositis.

5   Clinical Manifestations

The usual patient with dermatomyositis is a young- or middle-aged individual who subacutely develops symmetric proximal limb weakness. The patient also has a characteristic rash of the cheeks and eyelids, called the heliotrope rash, which is pinkish purple, and often affects the face, eyelids, or hands. There can also be red, scaling papules on the knuckles, sometimes called Gottron's papules. When disease is acute or severe, there can be muscle pain and tenderness. Clinical suspicion for dermatomyositis is confirmed by elevation of the muscle enzyme, creatine kinase (CK), in the blood, as well as by characteristic findings on electromyography. Some patients also have involvement of muscles of swallowing, leading to difficulty with swallowing, also called dysphagia. A substantial percentage of myositis patients have pulmonary involvement, called interstitial lung disease (ILD), which in some studies has identified a population of patients with a worse prognosis. The inflammatory myopathies can also be associated with other inflammatory diseases such as systemic lupus erythematosus (SLE), rheumatoid arthritis (RA), or Sjogren's syndrome.

6   Natural History and Prognosis

The natural history and prognosis of this disease are highly variable [1]. Most patients do poorly without treatment, and prior to the development of corticosteroid therapy the mortality rate was as high as 50%. Corticosteroid therapy has greatly alleviated the disease burden of dermatomyositis and other inflammatory myopathies.

7   Classification

The inflammatory myopathies are usually divided into dermatomyositis, polymyositis, and IBM. Most neurologists think of polymyositis as being identical to dermatomyositis, except without skin manifestations. This is reasonable, given the similarities in clinical presentation and in therapy. However, there are differences on muscle biopsy, and dermatomyositis is more frequently associated with malignancies. In contrast, IBM appears to be a somewhat different disease: the weakness is more slowly evolving, selectively involves finger flexors or quadriceps muscles, and usually appears in an older individual. CK levels may be normal in IBM, there are usually no associated autoantibodies, and it is not generally responsive to corticosteroid therapy. IBM also has characteristic biopsy findings not commonly found in the other two diseases. Some neuromuscular neurologists consider IBM to be more of a degenerative disease with the inflammatory response being secondary. However, recently Chahin and Engel [2] have described patients who had muscle biopsies characteristic of polymyositis but had clinical features consistent with IBM. Because of these overlap syndromes and the importance of determining response to immunomodulatory therapy, there have been increasing efforts over the past decade to classify these diseases by their molecular signatures.

8   Diagnosis

The diagnoses of polymyositis, dermatomyositis, or IBM are made on the basis of the characteristic symptoms and signs, elevated muscle enzymes in the blood, a consistent pattern on electromyography, inflammation on the muscle biopsy, and absence of other causes of myopathies or other associated inflammatory diseases, such as lupus, rheumatoid arthritis, mixed connective tissue disease, or vasculitis. It is important to distinguish the inflammatory myopathies from noninflammatory myopathies such as dystrophies, metabolic myopathies especially hypokalemia and thyroid disease, or muscle disease due to mitochondrial abnormalities, since treatment of these conditions is different.

Some autoantibodies associated with inflammatory muscle disease target a class of enzymes called aminoacyl-tRNA synthetases, which link a specific amino acid to transfer RNA. Antibodies to histidyl tRNA synthetases, also called anti-Jo-1, are the most common, but these are only seen in about a ¼ of myositis patients, while antibodies to the other aminoacyl transferases are seen in only 1–5% of patients. These autoantibodies, therefore, are not helpful clinically.

9   Therapy

The therapy most commonly used in dermatomyositis is corticosteroids, especially oral prednisone. The usual dose for initiation of therapy is 60 mg/day which can be adjusted over time depending on response to therapy or development of adverse effects. Corticosteroids have been very effective in the treatment of dermatomyositis and polymyositis. Therapy of IBM is more problematic; no therapy has been shown to be consistently helpful.

10   Related Neuroimmunological Muscle Diseases

(a) Polymyalgia rheumatica (PMR) is a relatively common inflammatory disease of unknown etiology, causing severe pain in muscles, usually in older individuals. Weakness is not a common feature. Usually, the pain is in the neck, shoulders, or hips and is worse in the morning. The diagnosis is made by the characteristic history, absence of other pathology, and confirmed by the presence of a high

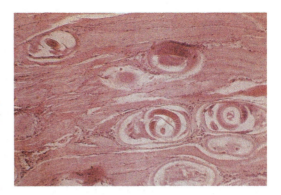

Fig. 11.1 Trichinosis. Histologic section shows encysted larvae of *T. spiralis* in skeletal muscle

erythrocyte sedimentation rate (ESR), a test of the blood which usually indicates an inflammatory process. In about 10% of PMR patients, a blood vessel inflammation called temporal arteritis, also called giant cell arteritis, occurs. This can be a dangerous condition, because it can occlude vessels to the eye, and cause blindness. PMR and temporal arteritis are highly responsive to corticosteroid therapy and should be high on the list of the diagnostic possibilities of older patients with persistent muscle pain.

(b) Diseases of the neuromuscular junction: Other neuroimmunological muscle diseases to be considered in the differential of patients with IIMs are neuromuscular junction diseases. Weakness is a hallmark of these diseases, such as myasthenia gravis and Lambert–Eaton myasthenic syndrome; they are reviewed in Chap. 10. Usually, the differences on history and examination between these diseases and inflammatory muscle diseases are clear, but occasionally extensive laboratory studies are necessary to distinguish them. In addition, there are infectious causes of muscle disease, such as viruses (including HIV and influenza), parasites (trichinosis—shown in Fig. 11.1—toxoplasmosis and cysticercosis), and even bacteria (such as staphylococcus). These are relatively rare.

11   Experimental Models of Inflammatory Muscle Disease

A variety of animal models of IIMs have been induced by immunization of experimental animals, usually rats, with muscle homogenates or purified components of muscle [3, 4] In this model, called experimental autoimmune myositis (EAM), investigators have used a variety of skeletal muscle proteins, including the C-protein, laminin, and histidyl-tRNA synthetase. Most of the publications on the models represent "proof-of-principle" types of experiments in which immunizations with various components were utilized simply to establish that these proteins could function as autoantigens to induce autoimmunity and injury. Within the past few years, some investigators have moved more aggressively into mouse models, in which mutant mice can be used to identify possible pathogenetic mechanisms. However, as of 2010, experimental models of inflammatory muscle disease have not contributed significantly to understanding the pathogenesis of these diseases.

References

1. Mammen AL. Dermatomyositis and polymyositis: Clinical presentation, autoantibodies, and pathogenesis. Ann N Y Acad Sci. 2010;1184:134–53.
2. Chahin N, Engel AG. Correlation of muscle biopsy, clinical course, and outcome in PM and sporadic IBM. Neurology. 2008;70(6):418–24.
3. Nagaraju K, Plotz PH. Animal models of myositis. Rheum Dis Clin North Am. 2002;28(4):917–33.
4. Katsumata Y, Ascherman DP. Animal models in myositis. Curr Opin Rheumatol. 2008;20(6):681–5.

A Introduction to Neurological Infections: Neuro-infectious Disease as Part of Neuroimmunology

12

Neurological infections are an important part of neuroimmunology. The concepts and issues dealt with in the previous chapters have covered neurological inflammatory diseases that are considered "autoimmune" primarily because no pathogen has yet been identified. These diseases also have relevance to neurological inflammatory diseases in which pathogens have been identified. The nervous system has only a limited repertoire with respect to inflammation. Processes such as microglial proliferation, loss of blood–brain barrier function, immune cell infiltration into the CNS, have already been discussed and function similarly in neurological infections as they do in autoimmune diseases.

In no field of medicine is the statement, "Absence of evidence is not evidence of absence" more relevant than in neuroimmunology and neuro-infectious disease. The likelihood that many neuro-inflammatory syndromes such as MS and neurosarcoidosis are caused by pathogens seems high and the fact that pathogens have not yet been identified is not strong "evidence of absence" since our current tools for identifying new infections in the CNS are not powerful. A recent example of this difficulty is Lyme disease. When Lyme arthritis was first described, many rheumatologists were convinced that it was autoimmune since the pathology of the joints mimicked the pathology of rheumatoid arthritis, all standard cultures of the joints were negative, and no organisms were seen on joint pathology. Initially, the disease was linked to ticks by epidemiological studies, and subsequently the spirochete was serendipitously identified in ticks from Lyme-endemic areas by Dr. Willy Burgdorfer, an entomologist. Serum from Lyme arthritis patients was then shown to bind to these spirochetes, and the spirochetal etiology confirmed in 1981 [1]. Since the rash of Lyme disease was first described by the Swedish dermatologist Afzelius in 1909, it took 72 years and luck to finally identify it as an infectious disease. Thus, many neuro-inflammatory syndromes which we now consider "autoimmune" or "inflammatory" may be infectious due to as yet unidentified pathogens. Their discovery as infections may simply be waiting for a serendipitous discovery similar to that of Dr. Burgdorfer.

Neurological infections can be classified in many ways, including by:
- Location within the nervous system (central nervous system—brain meninges, spinal cord parenchymal vs. peripheral nervous system)
- Pace (acute/subacute/chronic), or
- Type of infectious agent (bacterial, viral, fungal, and other).

This primer will utilize location, and, as in previous chapters, a prototypic disease will be discussed for each of these: Lyme disease/neurosyphilis for involvement of meninges, herpes simplex encephalitis (HSE) for involvement of the "encephalon" (enkephalos in Greek means brain, literally "within the head"), and leprosy for infections involving the peripheral nervous system. Other infections important for the neuroimmunologist will be discussed with these as a backbone.

A.R. Pachner, *A Primer of Neuroimmunological Disease*,
DOI 10.1007/978-1-4614-2188-7_12, © Springer Science+Business Media, LLC 2012

1   General Rules of Neuro-infectious Diseases

1.1   Infection of the Nervous System Follows Systemic Infection

Although infection sometimes appears first most dramatically in the nervous system and is what brings the patient to the attention of physicians, neurological infection is almost always a consequence of initial entry of the infection elsewhere, and subsequent spread to the nervous system. Frequently, in very ill hospitalized patients, bacterial meningitis is a consequence of sepsis associated with known infections such as pneumonias, abscesses, or urinary tract infection. Severe infections contiguous to the CNS, such as sinusitis or otitis, can spread to the CNS. In CNS tuberculosis, CNS infection follows pulmonary involvement, and in Lyme disease, neurological involvement follows the dermatological process. Frequently, though, in acute bacterial meningitis in the community, meningitis is the first sign of the infection, although clearly the infection was acquired systemically. The only true exception to this rule is after neurosurgical intervention or penetrating injuries to the CNS when bacteria are directly introduced (see Inset 12.1).

Inset 12.1 " A 45-Year-Old Man with a Brain Abscess"

JT was admitted to the hospital from the ER with a seizure. Diagnosed with paranoid schizophrenia in his late teens, he had been in and out of mental institutions for most of his adult life, but had had been doing fairly well at home until a month prior to his admission when he refused to take his medications because they "limited me." For the week before admission, his parents with whom he lived had noted that he became increasingly agitated. On the day of admission, intermittent loud sounds were coming from his upstairs bed room and when his father investigated he found his son unresponsive, in a pool of blood, after drilling

holes into his skull with an electric drill to "let the demons escape." Three holes had already been drilled and a seizure had been responsible for the loud sounds and the unresponsiveness. After 3 weeks of admission to the neurosurgical unit for management of his wounds, and treatment with anticonvulsants, antipsychotics, and antibiotics, the patient began complaining of worsening headaches and a CT scan revealed two abscesses which required excision.

Author's note. This patient is an exception to the rule that systemic infection precedes CNS infection. Usually, the penetrating head trauma is from motor vehicle accidents. In this case, the patient was a victim of self-injurious behavior.

1.2   Neuro-infectious Diseases Should Be in the Differential Diagnosis in Many Patients with Difficult Neurological Syndromes Because of Their Reversibility with Antibiotics

In the USA, Europe, and other countries with a high level of public health and hygiene, CNS infections are not as common as in developing countries. Thus, neurologists frequently do not consider CNS infections in the differential diagnosis of neurological conditions despite the fact that most infections are much more effectively treated than many of the other diseases neurologists diagnose. This is illustrated in the patient described in Inset 12.2, in which CNS toxoplasmosis was not considered in the differential diagnosis of a young man with right-sided weakness.

1.3   Diagnosis by Identification of the Infectious Agent Is Ideal, but Sometimes Impossible

Ideally, diagnosis in an infectious disease is made by isolation of the pathogen from the patient. For example, a bacterial meningitis is diagnosed by culture of the bacterium from the cerebrospinal

A 31-year-old construction worker, married and father of two, presented to a community neurologist in Newark, NJ, with 4 months of malaise and mild right-sided weakness. He had had some difficulty with vision 3 years previously which had resolved; he could not recall if this involved one eye or both. Examination confirmed mild right proximal weakness with mild hyper-reflexia on the right and a right Babinski; there were no other abnormalities. MRI of the brain revealed a number of enhancing lesions, and the diagnosis of multiple sclerosis was made; the patient was offered IFN-β therapy. A lumbar puncture was not performed. The patient requested a second opinion and came to our University Hospital. The history and examination was as previously obtained except that the patient admitted to the use of occasional intravenous drugs and promiscuous sexual activity over the previous 15 years. A review of the MRI revealed that the lesions were not typical of MS. An HIV antibody assay was positive with a high plasma viral load, and the CD4 count was low. Based on the brain MRI and the HIV positivity, a diagnosis of HIV/AIDS with CNS toxoplasmosis, infection of the brain with the parasite *Toxoplasma gondii*, was made and confirmed by a treatment trial, i.e., complete resolution of his headaches and MRI lesions by treatment with pyrimethamine and sulfadiazine for 3 weeks. He was subsequently placed on treatment with antiretroviral therapy (tenofovir/emtricitabine/efavirenz) with clearance of his plasma viral load. He did well in follow-up.

Author's note. Many urban areas in the USA have quite high incidences of HIV/AIDS. Neurologists practicing in the area must be aware of this problem and considers opportunistic infections in the differential diagnosis of neurological disease.

fluid (CSF) of the patient. Until recently, the diagnosis of viral infection by isolation of the causative virus was difficult, because of methodological problems with virus cultures. The advent of polymerase chain reaction (PCR) testing has been a major advance for diagnosis of viral infections of the CNS. For example, in the past the diagnosis of progressive multifocal leukoencephalopathy (PML), a demyelinating CNS disease seen in immunosupressed patients and caused by the JC virus, was only possible by brain biopsy. Now the diagnosis is confirmed, after suspicious clinical and MRI findings, by positive PCR for JC virus in the patient's CSF. Some organisms, such as *Mycobacterium tuberculosis*, the causative organism of tuberculosis, are notoriously difficult to diagnose by isolation of the offending organism, and organisms present in the CSF of infected patients can take weeks to grow in culture. In other infections such as Lyme disease and neurosyphilis, the problems with identification of the infectious agent, either by culture or PCR, are such that indirect methods, primarily identification of the specific antibody response, are used. Reliance on serology, however, makes the false-positive and false-negative rate in diagnosing these infections higher than for other CNS infections in which isolation of the infectious agent is straightforward (see "Patient with myasthenia gravis masquerading as Lyme disease" in Chap. 10).

1.4 Prompt Recognition of Curable Infectious Diseases Is Critical

It is axiomatic that curable infectious diseases of the nervous system need to be diagnosed and treated as soon as possible, but that is frequently easier said than done, especially if the index of suspicion for infection is low. Given the fact that infection incidence is dependent on geographic area of exposure, a patient's history of travel to regions far away from home is important. However, a travel history is frequently not obtained by health care professionals, an omission in our highly mobile society that can result in an unfortunate delay in diagnosis and treatment. What is common in one area is very rare in another, and physicians generally think about diseases likely in their specific region not in other areas (see Inset 12.3).

Inset 12.3 "Headaches and Confusion in a Surgeon in New Jersey"

In the fall, this 51-year-old orthopedic surgeon from NJ, father of three, developed headaches and mild confusion which forced him to cut back on his practice. When his confusion worsened, he saw his primary care physician who drew blood work and referred him to a local neurologist. Both physicians felt Lyme disease was possible, and he was begun on antibiotics for this infection. However, there was no history of a skin rash and a Lyme blood test was negative. When symptoms worsened, an MRI of the brain and an LP was performed. The MRI revealed numerous, but nonspecific white matter changes, and the lumbar puncture revealed mildly low glucose, elevated protein, and many white cells, mostly monocytes. Lyme antibody testing in the CSF and serum was negative. Because of the white matter changes, the neurologist diagnosed probable CNS vasculitis and instituted high-dose corticosteroid therapy. The patient was less confused and generally felt better for a few days, but then became agitated, belligerent, confused, and then drowsy and poorly responsive. After transfer to a large metropolitan center and subsequent meningeal biopsy, the patient deteriorated further and died. The biopsy and CSF revealed culture positivity for *Coccidioides immitis*, the serum was positive for anti-Coccidiodal antibodies, and autopsy revealed disseminated fungal lesions. The wife provided the history, previously not obtained, that her husband had spent 2 weeks in a tennis camp in Arizona in the late spring of that year. It was felt by the clinicians caring for him that earlier treatment, in conjunction with avoidance of corticosteroids, could have been curative.

Author's note. The initial diagnosis of possible Lyme disease was not unreasonable, given

that the patient was from NJ, a state highly endemic for that infection. However, the negative tests for anti-*B. burgdorferi* antibody essentially ruled out that process, which the community neurologist realized. His alternative diagnosis, primary CNS vasculitis, was highly unlikely, given its rarity (see Sect. 5 in Chap. 13). Thus, initiating corticosteroid therapy, which is an immunosuppressive, without more evidence for primary CNS vasculitis, an extremely rare disease, was quite risky, especially without having thoroughly ruled out an infection. Coccidioidomycosis, caused by *Coccidioides immites* (Fig. 12.1), is a relatively common infectious cause of subacute meningitis in Arizona but exposure does not occur in the Northeast and Midwest USA, while Lyme disease is a common infectious cause of subacute meningitis in the Northeast and Midwest USA, but exposure does not occur in Arizona.

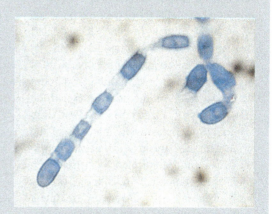

Fig. 12.1 *Coccidioides immitis.* Arthrospores (mycelia) of *Coccidioides* in culture. This mycelial phase of the organism is the form which grows in desert soil and in the laboratory. These spores easily become aerosolized and are infectious; it is these spores which get inhaled and cause the disease. In the mammalian host, the fungus changes into another form, a spherule

1.5 Therapy of Neuro-infectious Diseases Is Frequently More Difficult than Therapy of Infections Elsewhere in the Body

As in many situations in which inherent in a system's unique strengths are also unique weaknesses, in the CNS the isolation from the systemic circulation provided by the blood–brain barrier, which protects the CNS most of the time, also results in difficulties when trying to deliver anti-infective products to the CNS. Thus, most drugs injected intravenously only partially penetrate into the CNS. This is true even when the blood–brain barrier is partly opened by the inflammation associated with the infection. This difficulty with delivery of antibiotics to the CNS applies to treatment of any infectious process within the CNS. Sometimes if the infection is life-threatening and response to intravenously administered anti-infectives is insufficient, the physician is forced to deliver them directly into the subarachnoid space using intrathecal injections.

1.6 Always Keep HIV/AIDS in Mind

Over a million individuals in the USA are infected with human immunodeficiency virus (HIV) of which a large percentage are not aware of their infection. In addition, a significant percentage of individuals do not volunteer the fact that they are HIV-positive to health care professionals. Thus, one must always consider the possibility that a patient with a newly acquired neuro-inflammatory syndrome may have HIV/AIDS underlying their illness, since this diagnosis increases the possibilities of a large number of infectious diseases not normally encountered in the immunocompetent patient.

This book is not a neuro-infectious disease text; there are some excellent books which fill that bill [2, 3]. Instead, relatively common, prototypic neuro-infectious diseases will be described. As in earlier chapters, the diseases will be discussed by anatomic locations. Since most infections first encounter the nervous system via the meninges, meningitis will be the first topic.

The subacute spirochetal meningitides (neuro-syphilis/Lyme disease) will be discussed with brief reviews of acute bacterial meningitides and chronic meningitides. Some infections more commonly develop in the brain parenchyma, and HSE will be described as a prototypic disease for such CNS infections. Next, leprosy will be discussed as the prototypic peripheral nerve infection. Finally, neurological infections characteristic of HIV/AIDS will be discussed.

2 Lyme Meningitis: The Prototypic Meningitis

Most infections reach the CNS via the blood, and the blood vessels in the CNS traverse the meninges for a substantial distance before finally penetrating into the CNS parenchyma. The complex anatomy of the meninges is shown in Fig. 12.2. CSF circulates in the subarachnoid space, below the arachnoid and above the pia mater, in the space labeled "subarachnoid space." Pathogens causing meningitis frequently are unable to pass through the pia mater, and the glial limitans, to enter the brain parenchyma in any large numbers and thus remain in the subarachnoid space. Thus, it is not suprising that meningitis is the most common form of neurological infection. Almost any microorganism can cause a meningitis.

2.1 Definition

Lyme neuroborreliosis (LNB) results from infection of the nervous system by the spirochete *Borrelia burgdorferi*.

2.2 Etiopathogenesis

Spirochetes are ideally suited for persistence in the human host. Although they are very fragile outside of an animal, once inside they can survive for prolonged periods. Unlike viruses, which depend on rapid replication after entry into cells, spirochetes are extracellular and have very slow

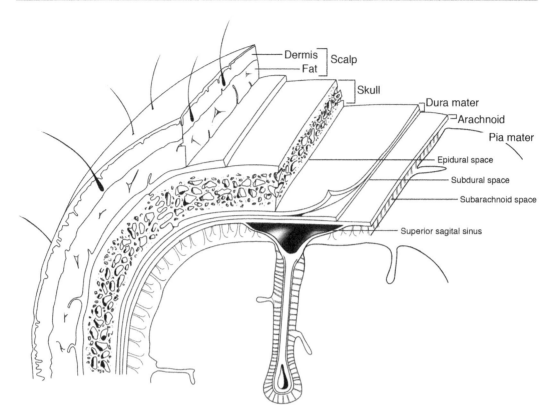

Fig. 12.2 Layers of the scalp, skull, and meninges. Intracranially, the space between the skull and the dura is the epidural space, the space between the dura and the arachnoid membrane is the subdural space, and the space between the arachnoid membrane and the pia membrane closely applied to the brain surface is the subarachnoid space. Cerebrospinal fluid (CSF) fills the subarachnoid space

rates of replication. Using a variety of means to escape the immune system, they are able to survive despite strong anti-spirochetal immune responses. They are large microorganisms with an associated large genome. *Borrelia burgdorferi* has a genome of 1.4 million base pairs, while Theiler's virus, which induces a chronic infection resembling MS in mice (see Chap. 8), in contrast has a genome of only 8.1 thousand base pairs. Only a very small percentage of *Borrelia burgdorferi*'s genome encodes for proteins which are characterized; the spirochete's genome thus represents a vast uncharted wilderness for molecular biologists. The most common spirochetal infections in the USA and Europe are Lyme disease, caused by *Borrelia burgdorferi*, and syphilis, caused by *Treponema pallidum*. Spirochetes from some types of clinical specimens can be readily seen using dark-field microscopy and appear as helical, motile organisms about 10–25 μm in length (Fig. 12.3).

There are a number of different subspecies of *Borrelia burgdorferi* which determine the type of neurological involvement in LNB. The clinical phenotype of LNB in humans and in experimental models is determined by the genotype of the infecting strain [4]. The whole species of *Borrelia burgdorferi* is called *Borrelia burgdorferi* sensu lato; LNB in North America is caused by the sensu stricto strain of *Borrelia burgdorferi*, the predominant form in North America, while LNB in Europe is caused by either *Borrelia garinii* or *Borrelia afzelii*. In Lyme disease, spirochetes can lodge anywhere where there is collagen and connective tissue. These can include collagen within the extracellular matrix in the meninges,

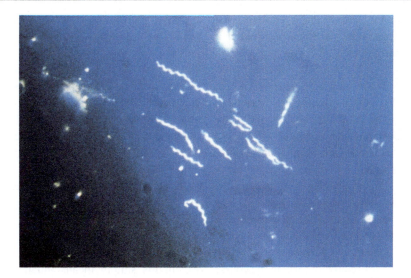

Fig. 12.3 *Borrelia burgdorferi*, dark-field microscopy. The spirochetes are corkscrew-shaped and are 10–20 μm long (from Centers for Disease Control and Prevention's Public Health Image Library (PHIL), with identification number #6631)

where fibroblasts are present, or in the connective tissue around nerves, the perineurium. Thus, these pathogens do not invade cells but survive around them. They may utilize the linear fibrils of collagen as a "hiding" place, to avoid immunosurveillance by lymphoid cells [5]. They also avoid immune-mediated destruction by interfering with host complement. The inflammation caused by these spirochetes thus tends to occur in the meninges and in peripheral nerves; neurons and oligodendroglia are not generally directly affected and thus permanent neurological dysfunction or demyelination is generally not a feature of LNB.

Autoantibodies, especially in the CSF, appear commonly in LNB and other chronic infections. In one study, the frequency of intrathecal synthesis of autoantibodies was 85% in LNB, compared with 12% in multiple sclerosis and 55% in viral meningo-encephalitis; intrathecal antibodies had specificity for more than 20 different CNS proteins, including myelin basic protein (MBP) and three neurofilament proteins [6]. Whether autoantibodies contribute to pathogenesis in this and other chronic infectious disease is unknown. The overlap between chronic infection, autoimmunity, and chronic inflammation is a recurrent motif among chronic infections.

2.3 Clinical Manifestations

Lyme meningitis is the most common manifestation of LNB in the USA and usually presents a few weeks to a few months after initial infection. Patients experience symptoms of headaches, neck stiffness, numbness, and tingling without clear anatomical localization, muscle aches, fatigue, and malaise; these symptoms are generally mild relative to the severe symptoms of meningitis caused by more aggressive bacteria such as meningococcal or streptococcal meningitis. Sometimes, patients develop focal neuropathies or radiculopathies, but severe neuropathies with substantial motor weakness are unusual. Many patients early in the infection think they are simply depressed or overworked or sleep-deprived, but later become more symptomatic, and come to the attention of the physician. Many patients have mild cognitive symptoms but not severe organic brain syndromes. Almost all patients with Lyme meningitis have had the rash of Lyme disease, erythema migrans, prior to the meningitis, but frequently did not notice it or ignored it after it resolved spontaneously. In European LNB, caused by strains of *B. burgdorferi* different from those causing American LNB, patients can, in addition to the chronic meningitis discussed above, also get painful inflammation of nerve roots, called Bannwarth's syndrome. These patients have radicular pain mimicking at times cervical or lumbar disk pain. Cranial nerve involvement is common in both American and European LNB. In European LNB, almost any cranial nerve can be affected, while in American LNB, unilateral or bilateral facial palsy is almost always the cranial nerve involved. The facial palsy usually occurs relatively early in the course of the infection, often early enough not to be associated with a substantial CSF pleocytosis, and almost never results in permanent significant deficit [7]. The triad of meningitis, cranial neuritis, and radiculoneuritis is not always present in LNB, but when it is, it is a distinctive clinical picture that may allow the infection to be diagnosed on clinical grounds [8].

The subacute/chronic meningitis of LNB is similar to other subacute/chronic meningitides caused by other infectious or inflammatory processes such as tuberculosis, fungal infections such as cryptococcus, sarcoidosis, and cancer. Lyme meningitis and the other causes of subacute/chronic meningitis listed above can cause either very chronic presentations lasting months or very acute presentations. Most meningitides caused by pyogenic bacteria such as staphylococcus or streptococcus are acute and patients become ill very rapidly; however, some bacteria such as *Listeria monocytogenes* can cause more chronic presentations.

2.4   Natural History/Prognosis

The natural history of Lyme meningitis is not known, since it is readily treated with antibiotics, once identified. The spirochete can be cleared and the symptoms resolved. Lyme meningitis is not fatal. Some patients are not identified until many months to years after onset of symptoms,

and these patients do not have a high incidence of major irreversible neurological sequelae. A small subset of patients who develop Lyme disease in Europe develop chronic CNS inflammation, such as myelitis, but this has not been seen with any regularity in American LNB. Thus, this infection can be chronic, persistent, but is frequently not damaging, even if not immediately identified. This is consistent with its proclivity to infect collagen-containing structures in the nervous system, rather than attacking neurons or glia.

2.5   Diagnosis

Meningitis simply means inflammation of the meninges, and the diagnosis of any meningitis is made by a constellation of symptoms, signs, and the demonstration of inflammation within the subarachnoid space as determined by analysis of the CSF after lumbar puncture. Once the diagnosis of meningitis is made, the causative pathogen/process must be identified. The most important immediate concern is meningitis caused by a pyogenic bacterium such as meningococcus, streptococcus, or staphylococcus, since these organisms can cause great damage quickly. Thus, gram staining of the CSF and bacterial cultures are essential in the acute management of the meningitis patient. Latex agglutination tests which rapidly identify antigens of the common bacteria of meningitis, usually *Streptococcus pneumoniae*, Group B streptococcus, *E. coli*, meningococcus (*Neisseria meningitidis*), and *Hemophilus influenzae*, have been used as a helpful adjunct in diagnosis, although their sensitivity and specificity are still uncertain. In Lyme meningitis, these tests are negative, because the spirochete which causes the disease is not culturable in routine bacterial media, and no latex agglutination assay is available. For Lyme meningitis, the laboratory diagnosis must be made from the detection of anti-*B. burgdorferi* antibody in the serum and CSF and by a history of the skin rash. Since high levels of IgG in the CSF can cause a false-positive CSF antibody result, an index is frequently helpful. The demonstration of intrathecal antibody production by measuring CSF and serum anti-*Borrelia* antibodies is the laboratory diagnostic tool of choice in many countries [9]. An explanation of CSF/serum testing for intrathecal antibody production is provided in Chap. 18. Many clinicians will diagnose Lyme meningitis based on an inflammatory CSF, negative bacterial cultures, and positive antibodies to *Borrelia burgdorferi* in the blood, although this approach can result in false-positive diagnoses when other pathogens are responsible in patients from an area endemic for *Borrelia burgdorferi* infection.

2.6 Therapy

Spirochetes are very sensitive to a variety of antibiotics including penicillin. However, the ability of many antibiotics, including penicillin, to kill the bacteria requires the bacteria to reproduce, and the doubling time of spirochetes in vivo is quite long. Therefore, most treatment regimens for LNB require a minimum of 2 weeks of therapy. Some clinicians insist on extensive courses of intravenous antibiotic therapies, despite evidence that a 2-week course is effective [10]. In fact, studies from Europe demonstrate that oral therapy of LNB with the oral antibiotic doxycycline is just as effective as intravenous therapy [11]. Some patients with LNB also have myalgias and arthralgias, which can benefit from therapy with nonsteroidal anti-inflammatory drugs (NSAIDs), in addition to the antibiotics.

2.7 Classification

(a) *By course*. Acute vs. subacute/chronic. The duration of Lyme meningitis is usually many days to a few weeks, and thus can be classified as a subacute to chronic meningitis, which contrasts it to acute meningitides such as viral meningitis/aseptic meningitis or pyogenic bacterial meningitis which develop over hours.

(b) *By CSF profile*. "Bacterial" vs. "aseptic." Patients with acute bacterial meningitis caused by pyogenic organisms such as pneumococcus

or streptococcus have dramatic CSF findings. Usually cloudy or in some cases even milky and thick, there are hundreds to thousands of white cells, mostly polymorphonuclear leukocytes (PMNs), associated with the presence of bacteria when special stains such as the Gram stain are used. The glucose concentration tends to be low and the protein high. The "aseptic" picture is very different: clear to slightly cloudy fluid, tens to hundreds of white cells, mostly lymphocytes or monocytes, with a normal glucose, mildly elevated protein, and absence of bacteria on Gram stain. Some bacteria, such as *Borrelia burgdorferi*, *Treponema pallidum*, or *Listeria*, are frequently not "pyogenic," i.e., pus-causing, and thus cause an "aseptic" picture. Thus, an "aseptic" pattern on CSF analysis does not rule out bacteria as the cause of the infection.

(c) *By pathogenic organism*. Pyogenic bacterial, spirochetal, viral, fungal, other (mycobacterial, rickettsial, parasitic, etc.). Most clinicians prefer to identify meningitides by their causative organism, since there are features outside of the CNS that can identify them, e.g., facial palsy in Lyme meningitis, pulmonary infiltrate in coccidioidal meningitis.

2.8 Selected Important Infectious Meningitides

2.8.1 Pyogenic (Bacterial) Meningitis

The most common causes of acute, non-spirochetal bacterial meningitis in adults are *Neisseria meningitidis* (meningococcus) and *Streptococcus pneumoniae* (pneumococcus). Almost all patients present with two of the four following symptoms: headache, fever, neck stiffness, and altered mental status [12]. In patients over the age of 50, *Listeria monocytogenes* is an increasingly common cause of acute meningitis. Bacterial meningitis is a medical emergency, which needs to be diagnosed and treated as quickly as possible by lumbar puncture and intravenous antibiotics. Delays in diagnosis and therapy can be fatal.

The immune response to bacterial meningitis is almost completely an innate immune response which is usually inadequate to control the rapid replication of the bacteria. Since the presentation is so acute, the more powerful adaptive immune response does not have enough time to become armed and active. Thus, the spinal fluid shows large numbers of PMNs, not lymphocytes; bacteria also usually can be seen in the CSF by a technique called the Gram stain. Since the infection causes problems with reabsorption of CSF, the CSF is also frequently under increased pressure.

2.8.2 Cryptococcal Meningitis

This form of meningitis is especially important because it is one of the more common causes of meningitis in patients with HIV/AIDS [13]; it is also diagnosed in immunocompetent individuals. The pathogen of cryptococcal meningitis is *Cryptococcus neoformans*, a fungus which has a characteristic gelatinous outer covering. This characteristic makes it easily detectable in CSF if India ink is added since the background becomes black and the fungus with its gelatinous coat becomes bright from reflected light (Fig. 12.4). It can also be tested for antigenically using the cryptococcal antigen test or directly cultured using fungal cultures.

Its clinical manifestations are similar to those of Lyme meningitis outlined above, with chronic headache, malaise, and fatigue; fever is more commonly present, but may be low grade. Unlike Lyme meningitis, cryptococcal meningitis can cause raised intracranial pressure, due to decreased reabsorption of CSF and decreased level of consciousness. Also, whereas Lyme meningitis usually affects only the facial nerve, cryptococcal meningitis can cause impairment of almost any cranial nerve. When it occurs in patients who are HIV-positive, especially if the CD4 count is less than $100/mm^3$, the clinical course may be rapidly downhill; mortality of the infection under these circumstances is high even with aggressive treatment. Treatment usually consists of the antifungal agents amphotericin B and flucytosine.

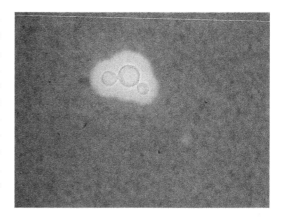

Fig. 12.4 Encapsulated budding yeast cells of *Cryptococcus neoformans* in cerebrospinal fluid delineated by India ink particles

2.8.3 Syphilitic Meningitis

Syphilitic meningitis is one of the presentations of neurosyphilis, which is neurological involvement with the spirochete *Treponema pallidum* and resembles Lyme meningitis in many ways. Neurosyphilis is discussed later in this chapter under diseases associated with HIV/AIDS.

2.8.4 Tuberculous Meningitis

TB meningitis is one of the most common forms of meningitis in the world, since *Mycobacterium tuberculosis* infects one-third of the world's population, but its incidence in the USA is low. For instance, the yearly incidence of new cases of active TB in the USA is 4/100,000 compared to the incidence in Swaziland in Africa of 1,200/100,000. TB meningitis is more common in the pediatric population [14] and presents similar to Lyme meningitis with headache, stiff neck, cranial nerve involvement, malaise, fatigue; however, in TB meningitis, patients tend to appear more acutely ill with a higher incidence of fever and decreased level of consciousness. The presentation of TB meningitis in children is a bit different than in adults with a higher percentage of nausea and vomiting and of prior TB in the former. Very low CSF glucose (hypoglycorrhachia) is also a characteristic feature of TB meningitis. TB meningitis can also be a lingering chronic meningitis. In 83 cases of chronic meningitis in an immunocompetent population in New Zealand, the most common cause by far was tuberculosis in 60% [15]. Although TB treatment is generally fairly effective, the combination of an increasing prevalence of multidrug resistant bacteria plus the increasing concurrence of HIV-positivity has led to a higher than expected mortality rate for TB meningitis.

2.8.5 Aseptic Meningitis

The most common form of acute meningitis in the general population is viral meningitis caused by transient adenoviruses or other relatively mild viral infections. These are usually due to CNS seeding from a systemic viral infection. They cause headache, malaise, and sometimes mild neck stiffness for a few days, and then resolve without the need for therapy, as the systemic viral infection is cleared by the immune response. Occasionally, the symptoms become severe enough to prompt a visit to the ER resulting in a lumbar puncture which yields an "aseptic meningitis" picture and the patient is sent home with NSAIDs and the recommendation to maintain hydration and reduce fever. "Aseptic" which means "absence of sepsis" is an appropriate name, since the aseptic meningitis pattern on the CSF analysis is basically a compendium of the absence of features of bacterial meningitis, as discussed above. That is, there is no predominance of PMNs, there is an absence of raised pressure, and there are no bacteria. It is not unusual for an occasional patient with aseptic meningitis to be misdiagnosed as having bacterial meningitis, since the rules for identifying a CSF as "aseptic" are not inviolable, and clinicians will lean toward being conservative and overdiagnose, rather than underdiagnose, bacterial meningitis. Occasionally, if a lumbar puncture is performed in patients with very early bacterial meningitis, an "aseptic" picture can be present, and the patient can be inappropriately be diagnosed with viral meningitis.

3 Herpes Simplex Encephalitis: The Prototypic CNS Parenchymal Infection

Growing masses within the parenchyma of the brain, whether they are infections or cancers or other processes, cause symptoms primarily based on the size and the speed of their growth and their location within the CNS. The brain cannot expand beyond the limits of the skull, so space is at a premium. Very large growths can occur with minimal symptoms if they grow slowly and are not "irritating," i.e., do not cause inflammation or seizures. As an example frequently seen in practices with large numbers of immigrants from Mexico and South America, one parenchymal brain infection, neurocysticercosis, frequently results in many cysts throughout the brain often with no symptoms whatsoever. These cysts are sometimes discovered when patients complain of muscle aches and cysts are found in muscle; brain imaging is then obtained and shows multiple brain cysts without neurological symptoms. A similar situation can occur with the related parasitic infection, echinococcosis. These parasitic infections can be considered as abscesses since the CNS resident cells are not infected. In contrast, infections of the brain resident cells are called encephalitis, the pathogen is usually viral, and the clinical presentation is usually rapid and dramatic. The most common viral encephalitis in the USA and developed countries is HSE.

3.1 Definition

HSE is an acute, frequently fatal, encephalitis caused by the large DNA virus, herpes simplex virus (HSV).

3.2 Etiopathogenesis/Pathology/ Genetics/Epidemiology

The pathogen is a large and complex DNA virus and can be visualized in infected tissue by electron microscopy. On the outside of the virus are glycoprotein spikes, embedded in a lipid bilayer. Internal to the lipid bilayer is the tegument, a collection of proteins important for replication. Internal to this is the nucleocapsid, containing double-stranded DNA, as well as associated proteins.

Eighty to ninety percent of the population is infected with the HSV in an asymptomatic and latent form. The virus can reactivate, i.e., begin replicating in an infected host, either asymptomatically or symptomatically. With the latter, the symptoms are usually related to a characteristic sore in the lip or skin around the lips. At times when the virus is latent, it exists in the trigeminal ganglion and other ganglia just outside the brain.

Investigations of HSV infection in experimental animals has demonstrated that, when activation results in encephalitis, it begins in the trigeminal ganglion, rather than coming from blood-borne infection. Once within the brain, the virus lyses neurons rapidly and usually spreads from a temporal or frontal initial focus. The mechanisms by which virus penetrates the CNS, evades the anti-HSV immune response and lyses neurons in the brain are poorly understood. It is important to remember that HSE is an infection of neurons, as are most viral encephalitides, and thus a gray matter infection. Its clinical presentation is very different from infections of oligodendrocytes and white matter infections, such as PML (see below).

One of the best studied acute viral brain infection is that of mouse hepatitis virus (MHV) discussed briefly in Chap. 8. It is generally considered a model of MS because demyelination is prominent, but it is also an excellent model of acute viral encephalitis. Within the first few days of infection, innate immune defenses within the CNS, such as type 1 interferons (IFN-α and IFN-β), are critical to control the infection. Beginning after the first few days and extending to 2 or 3 weeks after infection, virus-specific cytotoxic CD8-positive T cells are active in preventing viral replication. To provide long-lasting prevention of viral growth, neutralizing antibody, defined as virus-specific antibody which can prevent virus infection of target cells in vitro, is produced by virus-specific plasma cells which begin to produce high levels of antibody after the first few weeks of infection. These antiviral weapons are chief among many used by the immune system in a sequential, coordinated manner to clear MHV from the CNS [16]. Similar mechanisms are operative in all encephalitides including HSE.

The vast majority of patients with HSE are not immunocompromised by standard criteria. Because HSE presumably results from a switch of virus from latency to reactivation, much interest has centered on a type of RNA called latency-associated transcripts (LATs), which are present in the ganglia of latently affected cells, interfere with viral killing, and prolong the survival of infected neurons. As would be expected, the pathology of HSE is dramatic with a combination of inflammation and necrosis, and loss of neurons. There are no clear genetic predispositions, although some recent research identifies mutations in genes controlling type 1 interferon responses as predisposing to HSE [17]. The virus and its encephalitis appear in nearly all populations. HSE is sporadic and causes about two cases per million population per year, so, for the vast majority of infected humans, HSV infection is either completely asymptomatic or only a minor occasional irritant because of transient "cold sores."

3.3   Clinical Manifestations

In most cases of HSE, the presentation is that of an acute focal cerebral lesion, usually localized to the frontal lobe or temporal lobe of one side. This is in distinction to most other viral encephalitides (see below) in which involvement of the brain is relatively diffuse, and the clinical syndrome is dominated by progressive confusion and decreased level of consciousness. The time course is hours to a few days. Headache and stiff neck are common symptoms for this as well as other meningoencephalitides. Many patients have acute cognitive difficulties, personality changes, or confusion and disorientation. Seizures are common. Fever is almost universal, and its absence raises questions about the accuracy of the diagnosis. Although imaging reveals a focal lesion, only about a third of patients manifest a clear focal abnormality on examination, with a hemiparesis or other focal finding. Patients can worsen quickly and HSE can be fatal (see Inset 12.4).

A 21-year-old college student had the worse headache of her life associated with fever and came to the emergency room in the early evening. She was evaluated initially by the ER staff who noted a temperature of 103.5, but her neck was not stiff. The patient's friends, who stayed with her in the ER, said that they thought she was a bit confused, but she was not disoriented according to the ER physician. She did not have a white cell elevation in her blood count, a CT scan of the brain was negative, and the neurology resident who saw her in the late evening made the diagnosis of possible aseptic (mild viral) meningitis, did not wish to wake the attending on call, so admitted her for possible further workup on the next day. When the attending made rounds on the admitted patients early the next morning, he found the patient unresponsive, with irregular respirations, with a third nerve palsy on the left (a "blown pupil") caused by pressure of the brain on the third cranial nerve. Over the night, the herpes virus had quickly grown in her brain, killed many brain cells, and the swelling from the cell death and inflammation caused pressure on her brain stem, a syndrome known as raised intracranial pressure (ICP), which, when very severe, causes herniation of the brain, i.e., protrusion of brain material through a variety of openings at the base of the skull. The patient underwent emergency surgery to decompress her brain and was treated with aciclovir, but died after a weeklong hospitalization. The autopsy confirmed the diagnosis of HSE.

Author's note. I rounded with the attending that morning on my first hour as a neurology resident; it was a frightening introduction to neurology.

3.4 Natural History and Prognosis

The natural history of the disease has been dramatically altered by the use of antiviral therapy. Prior to the use of antiviral therapy, the disease was fatal in the about 70% of patients, with the survivors usually having significant residual neurological disability. In two large studies in Sweden and the USA in the 1980s, aciclovir outperformed vidarabine and lowered mortality to 28% in one study and 19% in the other, with more than 50% of patients returning to normal life [18, 19]. The prognosis is best in those in whom aciclovir is begun soon after the onset of first symptoms.

3.5 Diagnosis

A high index of suspicion should always be present for the diagnosis of HSE, since it is the most common sporadic viral encephalitis, is treatable with antiviral medication, and the mortality for untreated HSE is very high. Thus, any individual with fever and clear neurological symptoms or signs such as headache or confusion or seizures should be immediately evaluated, and therapy initiated as soon as possible if there is any clinical suspicion of HSE.

Imaging of the brain is ideally performed by magnetic resonance which usually is clearly abnormal revealing a focal lesion unilaterally in the frontal or temporal lobe. Often there will be mass effect showing that the lesion is swollen and compressing other brain structures. Lumbar puncture reveals an inflammatory picture with large numbers of white cells sometimes associated with red blood cells.

The most sensitive and specific test for HSE in the acute clinical setting is the detection of HSV DNA in the CSF by PCR. Since the turn around time for this assay, i.e., the time from when the CSF is obtained to the time the result has returned from the lab, is usually about 12–48 h, treatment with antiviral medications should not be delayed waiting for the diagnosis to be confirmed. The PCR of the CSF for HSV is highly sensitive; thus, the absence of CSF PCR positivity is strong evidence against the diagnosis of HSE.

Fig. 12.5 Gertrude Elion was a researcher at Burroughs Welcome who received the Nobel Prize in 1988 for the invention or development of many drugs including aciclovir, 6-mercaptopurine, azathioprine, allopurinol, pyrimethamine, and trimethoprim (from the National Cancer Institute, http://commons.wikimedia.org/wiki/File:Nci-vol-8236-300_Gertrude_Elion.jpg)

3.6   Therapy

Aciclovir, also called acycloguanosine, is an effective drug to treat HSV infection, and has demonstrated efficacy in HSE. The standard dose is 10 mg/kg every 8 h for 10 days. The drug is phosphorylated to acycloGMP by viral thymidine kinase, and then to acycloGTP by cellular kinases. AcycloGTP is a potent inhibitor of viral DNA polymerase, which is the major enzyme responsible for viral replication. Aciclovir was one of many antibacterial and antiviral products invented by Gertrude Elion (see Fig. 12.5).

3.7   Other Important CNS Parenchymal Infections

3.7.1   Retroviral Infections of the CNS Parenchyma: HIV and TSP

Human Immunodeficiency Virus

The most common neurological syndrome in patients infected with HIV is HIV-associated neurocognitive disorder (HAND). The etiopathogenesis of this disorder appears to be direct viral infection of the brain, associated with varying degrees of host response to the virus although other effects cannot be completely ruled out [20]. The cells most infected with virus in the brain are those of the macrophage lineage, i.e., perivascular monocytes, macrophages and microglial cells, but astrocyte infection may also be important. HAND is usually classified as a subcortical dementia, in which concentration/attention, processing speed, executive function, learning and memory are most affected, with language and visuospatial processes less affected. There are no reliable assays or biomarkers, and the disorder is diagnosed by exclusion in an HIV-positive individual when other causes have been ruled out. In the past, CSF HIV levels had been a helpful adjunct of diagnosis, but recently, with the advent of effective antiretroviral therapies (ART), CSF HIV levels can be undetectable. Another benefit of effective ART has been the rarity of very severe dementia, which had been relatively common in the pretreatment era. Despite the clear benefit of ART early in infection with HIV, HAND may actually improve when stable patients discontinue ART, a surprising recent finding by HIV researchers, who had actually hypothesized the reverse when they began their study [21]. This finding led the authors to conclude that "the balance between the neurocognitive cost of untreated HIV viremia and the possible toxicities of ART require consideration."

The efficacy of ART also has another negative side in patients with opportunistic infections (OIs) associated with HIV, i.e., patients who are dually infected with HIV and an associated CNS pathogen such as cryptococcus, tuberculosis, or JC virus. Patients with OIs who have a combination of a high OI pathogen load and severe immunosuppression from their HIV may develop active inflammation in the CNS, if antipathogen immunity is restored by highly active retroviral therapy; this inflammatory syndrome is called immune reconstitution inflammatory syndrome (IRIS) [22]. IRIS is not a problem unique to the CNS and can be seen in other tissues, such as the eye in cytomegalovirus retinitis. The therapy of IRIS is difficult, but most use anti-inflammatory medications, despite their immunosuppressive properties. The peculiarities of PML-IRIS, which is being increasingly recognized, will be discussed below.

HTLV1-Associated Myelopathy/Tropical Spastic Paraparesis

HTLV1 is a retrovirus which exists as a provirus associated with T cells, and, unlike HIV, is not usually detected in plasma. The infection is found in endemic areas in sub-Saharan Africa, the Caribbean, South America, and Japan, where as many as 1% of individuals are infected with the virus. Of those infected, less than 5% ever develop any spinal cord involvement. HTLV1-associated myelopathy/tropical spastic paraparesis (HAM/TSP) is generally a very slowly progressive myelopathy with symmetrical weakness in the legs and less so in the arms. It also causes bladder symptoms. It is mildly inflammatory and causes atrophy over time. The diagnosis is made by exclusion of other spinal cord diseases in patients who test positive for anti-HTLV1 antibodies in the serum, and can be confirmed by demonstration of HTLV1 in peripheral blood mononuclear cells. There is no therapy that is generally accepted as successful.

3.7.2 Brain Abscesses

Whereas viruses cause CNS infections by actively infecting CNS cells and frequently lysing them, bacterial and parasitic infections, often occur in the CNS parenchyma as one or more abscesses, localized collections of infection which are usually walled off and cause symptoms by compressing normal CNS or by causing seizures. The most common bacterium is streptococcus, but any bacterium can cause a brain abscess. The cause is usually dissemination to the CNS from the blood and episodes of bacteremia usually precede symptoms by a few weeks, sometimes from a source of bacteremia as benign as skin pustule or a tooth cleaning. Parasites, such as *Toxoplasma gondii*, *Echinococcus*, or *Taenia solium* (causing cysticercosis), also can cause brain abscesses. Tuberculosis can cause abscesses that are sometimes called tuberculomas (Fig. 6.1). The diagnosis is sometimes difficult, but MRI scanning of the brain usually reveals characteristic findings of spheroidal masses with thick walls in patients with consistent histories and examination. In bacterial abscesses, surgical excision is frequently necessary, while

sometimes systemic antimicrobials are sufficient. An example of the clinical presentation of a patient with a brain abscess is in Inset 12.5.

Inset 12.5 A 38-Year-Old Policeman with Double Vision

A 38-year-old policeman over a few days noticed increasing problems with double vision, went to his primary care physician, who referred him to a local neurologist, who recommended immediate evaluation in our emergency room. History and examination was positive for a previously healthy man without risk factors who had a 4-day history of double vision and a 2-day history of mild headache. Patient recalled a dental procedure 1 month prior to symptoms, and examination was negative except for a marked right sixth nerve palsy, leading to double vision most marked on gaze to the right. Imaging of the brain revealed a large oblong mass above the pons which markedly enhanced after contrast. The patient was begun on antibiotics, taken to surgery, and a walled abscess removed which contained turbid fluid; the fluid grew in culture *Streptococcus viridans* and anaerobic organisms. After surgery, and a month of antibiotics, the patient progressively improved and eventually returned to work.

Author's note. This patient had a classical brain abscess, likely related to seeding of bacteria into his blood stream during the dental procedure. It is reported that approximately 10% of brain abscesses are due to bacteremia during dental procedures. However, this occurs so rarely related to the number of dental procedures that the risk of giving antibiotics as prophylaxis outweighs the risk in the vast majority of individuals. However, prophylactic antibiotics are given to some high-risk individuals such as those with certain cardiac conditions; our patient was a healthy man and did not receive prophylactic antibiotics during his dental procedure.

3.7.3 Progressive Multifocal Leukoencephalopathy

PML is caused by uncontrolled infection of the brain with a polyoma virus called JC virus. Neurological involvement occurs in only a very small percentage of those infected with this virus. The virus is very common as a subclinical infection in a large percentage, between 55 and 80%, of the general population. PML, an aggressive, frequently fatal CNS infection, has only been seen in immunosuppressed patients since immunocompetent individuals are able to control the JC virus. However, precisely what "control" means is unclear, since most highly immunosuppressed, JC virus-positive individuals do not get PML; thus, there are many host factors in the control of JC virus which we do not understand. PML was first described by Astrom, Mancall, and Richardson in 1958 when they described three patients, all with lymphoid malignancies, who developed "progressive multifocal" white matter lesions (leukoencephalopathy) and died in a few months after onset of neurological symptoms. Any immunosuppressed patient is susceptible to the disease, and, for many years, the leading underlying cause of immunosuppression leading to PML was AIDS. Now, an increasing number of cases of PML occur in patients receiving immunosuppressive therapy, including therapeutic monoclonal antibodies such as natalizumab, rituximab, and efalizumab, targeting the molecules α-4 integrin, CD20, and LFA-1, respectively. α-4 integrin and LFA-1 are primarily adhesion molecules and CD20 is a B-cell antigen. Because of PML cases, efalizumab, used for psoriasis, was withdrawn from the market, while natalizumab was initially withdrawn, and then returned to the market with substantial changes in how it was monitored. PML has also been seen in patients treated with mycophenolate.

The clinical presentation of PML is different than that of the prototypic CNS parenchymal infection HSE. Symptoms of PML are clinically more similar to those of acute attacks of MS. That is, brain white matter tracts, frequently many of them, are targetted (hence the name "multifocal"), but in PML there are no remissions as there are in MS, and the disease is generally progressive and fatal. The MRI frequently shows large confluent white matter lesions (Fig. 12.6), which progressively develop over time usually over a few months. Since the infection affects oligodendrocytes and white matter tracts, not neurons and gray matter, seizures are not common. PML patients can develop symptoms related to cortical involvement such as aphasias and visual field cuts. Patients in whom immunosuppression can be reversed sometimes survive. There is no effective antiviral therapy for JC virus.

PML-IRIS (see IRIS above) occurs when patients with PML undergo a change in therapy to reverse immunosuppression [23], and the now restored immune response aggressively attacks the JC virus in the brain, with associated inflammation and worsening of symptoms and CNS imaging. The change in therapy that can trigger PML-IRIS can be either anti-retroviral therapy in AIDS or removal of immunosuppressive therapies such as the immunosuppressive monoclonal antibodies. PML-IRIS resembles conceptually IRIS seen with other opportunistic infections such as CMV or cryptococcus in the setting of AIDS. PML-IRIS associated with monoclonal antibody therapy is still rare enough that clinical experience has not determined whether it behaves the same or somewhat differently from PML-IRIS associated with AIDS.

3.7.4 Other Viral Encephalitides: West Nile and Other Viral Encephalitides

West Nile virus

The West Nile virus (WNV) is a mosquito-borne RNA virus in the flavivirus family which causes encephalitis in humans and horses. Other viruses which cause human encephalitis in this family include St. Louis, Japanese, and Murray Valley encephalitis viruses. It was originally isolated from a human in the West Nile region of Uganda in 1937 and is part of a large group of encephalitis-causing viruses called arboviruses (*ar*thropod-*bo*rne *viruses*), which also include eastern and Western equine encephalitis. Insects other than mosquitoes can carry arboviruses and transmit encephalitis, such as ticks which carry tick-borne encephalitis (TBE). TBE causes thousands of

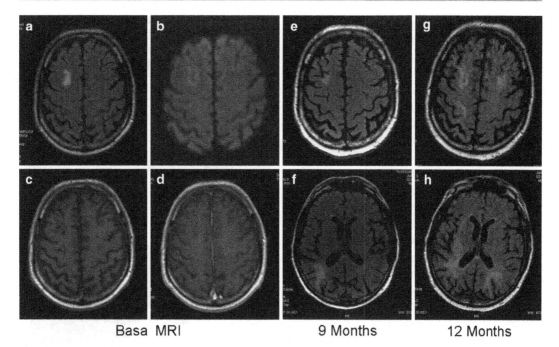

Basa MRI   9 Months   12 Months

Fig. 12.6 Large confluent white matter lesions of PML on brain MRI. The first MRI examination (**a**–**d**) revealed a left frontal homogeneous hyperintense lesion on axial fluid-attenuated inversion recovery (FLAIR) images (**a**) corresponding to a hypointense (**c**) nonenhancing (**d**) area on T1-weighted images. Diffusion-weighted imaging image (**b**) of the same lesion revealed a central core of hypointensity surrounded by a rim of high signal intensity suggesting a demyelinating lesion. Nine months later, the left frontal lesion showed a slight size reduction with atrophic feature of involved cerebral gyri (**e**). Nevertheless, similar white matter lesions involving also the subcortical "U" fibers with a "scalloped" appearance developed in the left temporal and parietal lobes (**f**). White matter lesions progressively became multiple and confluent, affecting asymmetrically both cerebral hemispheres as shown in axial FLAIR images obtained 12 months later (**g**–**h**)

cases of encephalitis in Russia and Europe annually. The tick vector of this infection, *Ixodes ricinus*, is not found in the USA and is also a vector of Lyme disease in Europe. Prior to 1999, WNV was found only outside of the USA and Europe, primarily in Africa and the Middle East. In the summer of 1999, an outbreak of WN encephalitis occurred in New York City. During subsequent years, the virus spread across the USA, becoming established in mosquito and bird populations. The epidemic in the USA in humans peaked in 2003 when there were almost 3,000 reported cases of WNV meningoencephalitis in the USA [24]. Since then the number of cases in the USA has been in the many hundreds per year.

The diagnosis of WNV encephalitis is made by identification of antibodies to WNV in the serum and either antibodies or viral nucleic acid in the CSF. This infection is distinctive because of two attributes. First, the isotype of immunoglobulin normally tested for is IgM since this isotype occurs at high titers and is persistent for months in this infection, in contrast to specific IgM antibody in most infections where IgM is a transient phenomenon of the first few weeks of the infection. Second, some patients with WNV infection can develop a myelitis characterized by acute flaccid weakness which has been called a poliomyelitis-like syndrome because it affects anterior horn cells.

The diagnosis of most encephalitides including WNV is made usually by clinical grounds when an epidemic of the virus is present. Otherwise, serology can be helpful in WNV infection especially rising IgM anti-WNV antibody levels; PCR or viral culture of the CSF can also be available. Therapy of most viral encephalitides is with supportive measures; antivirals have not been tested extensively in most non-herpetic viral encephalitides.

3.8   Rabies, Poliomyelitis, and the Wonder of Vaccination

Two parenchymal CNS infections, rabies and poliomyelitis, are debilitating infections which are now rare in the USA because of the effectiveness of the combination of reliable vaccines and stringent public health measures. Since 1999, there have been no cases of polio in the USA, while 50 years prior to 1999, in 1949, there were 42,033 reported cases. There are one to two cases of human rabies per year in the USA, compared to 55,000 deaths per year worldwide, mostly in Asia and Africa, according to WHO statistics. For polio, 10 years ago, there was hope that polio could be eradicated, similar to smallpox, after wide use of the oral vaccine had decreased the annual global incidence from 350,000 in 1988 to less than a thousand at the turn of the century. However, since 2000, there has been an increase in the number of polio cases, likely because 17% of the world's children are not vaccinated. This represents a failure, hopefully temporary, for the WHO, which has targetted polio as one of only two diseases for global eradication; the other is dracunculiasis, or Guinea worm disease, which does not have significant neurological sequelae.

Polio is caused by poliovirus, a picorna virus, i.e., small RNA virus, which is found naturally only in humans. Other neurologically important picornaviruses are Theiler's virus (see Chap. 8) and Vilyuisk encephalitis virus [25]. Despite the absence of an animal that can naturally be infected with poliovirus, research on poliovirus has been aided by the use of a mutant mouse model which expresses the human receptor for poliovirus (see Chap. 8 for more information on mutant mouse models) and develops poliovirus infection. Other picorna viruses causing disease outside of the nervous system in humans are enteroviruses, Coxsackie virus, and hepatitis A virus, among others. More than 99% of human infections with poliovirus are asymptomatic or result in only transient symptoms, but in some individuals the poliovirus invades the CNS, and specifically targets motor neurons, usually in the spinal cord. Most individuals with paralytic poliomyelitis survive the initial infection, but because the motor

Fig. 12.7 Children after polio, demonstrating the use of bracing to support weak and atrophied legs (from Lehava Netivot via the PikiWiki—Israel free image collection project http://commons.wikimedia.org/wiki/File:PikiWiki_Israel_)

neurons are destroyed, there is weakness and wasting of the affected areas and leg bracing is required to allow use of the legs for ambulation (Fig. 12.7). Poliomyelitis is an example of a disease which affects the spinal cord, but causes a predominantly lower motor neuron (LMN) presentation [see Chap. 2 for a description of LMN versus upper motor neuron (UMN) injuries], and in that way mimics amyotrophic lateral sclerosis. There are two types of polio vaccine: the oral Sabin vaccine and the injected Salk vaccine. Each has its advantages and disadvantages. The critical point is that since the infection is *only* found in humans, the disease can be eradicated globally if an adequate percentage of humans are vaccinated.

The situation is different for rabies. Rabies is a larger virus, and in a different family, i.e.,

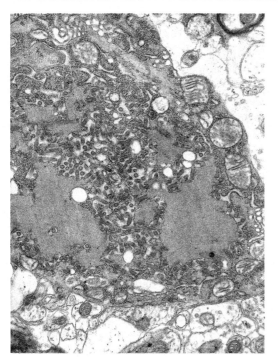

Fig. 12.8 Rabies virus. This is an electron micrograph of a cell at high magnification which shows *multiple bullet-shaped* rabies viruses within the cell (from the CDC, http://commons.wikimedia.org/wiki/File:Rabies_Virus_EM_PHIL_1876.JPG)

Rhabdoviruses (Fig. 12.8). The major reservoir of rabies virus is not humans but is animals, including skunks, raccoons, bats, foxes, dogs, and other animals. For each human case of rabies, there are thousands of animals infected with the virus. Thus, vaccination of a human will prevent infection in that human but will not affect the burden of the virus in the community, or the likelihood that others will be infected. For this reason, routine rabies vaccination is only recommended for individuals at higher risk of getting infected, such as animal handlers or those working in animal care clinics, shelters, or hospitals; in some cases international travelers going to countries highly endemic for the infection may consider rabies vaccination. Rabies in humans occurs after the bite of infected animals, or exposure to the saliva of infected animals and the presence of an open cut. By mechanisms poorly understood, the virus is thought to travel into peripheral nerves

and subsequently enters the CNS; some experiments in experimental animals have suggested that rabies may enter the CNS after being blood-borne, but this is not accepted as the usual mode of transmission in wild-type rabies. There is a fairly long incubation period from initial infection until entrance into the CNS, usually weeks to months, and during this time the infection can be treated by postexposure prophylaxis (PEP), consisting of wound cleansing, administration of rabies immune globulin, and rabies vaccination. However, because the incubation period is not highly predictable and can be shorter than expected, it is recommended that PEP begin as soon as possible after the bite. If PEP is not carried out, the virus makes its way to the CNS. The first symptoms of CNS invasion begin with flu-like symptoms—malaise, headache, and fever—progressing to encephalopathy, agitation, confusion, and finally delirium and coma. The disease is untreatable once the infection reaches this stage, and the very few individuals who have survived have had severe brain injury.

4 Infectious Neuritis: Neuropathies of Leprosy[1] and Varicella Zoster Virus

These neuropathies bear some resemblances to the prototypic neuroimmunological peripheral nervous system disease, Guillain–Barré syndrome (GBS), discussed in Chap. 9. They all cause neurological symptoms such as numbness, tingling, and weakness, by involving only nervous system structures outside of the central

---

[1] Some readers may question why I am including leprosy in this primer. Most physicians practicing in the USA or Europe will never see a patient with leprosy; the total number of reported cases in the USA in 2002 was 92. Leprosy is important for at least three reasons. First, it is a relatively common infection worldwide with millions of individuals disabled because of the infection. Second, it commonly affects the nervous system. Third, leprosy in the peripheral nervous system serves as a model of chronic infection and the interplay between helpful and harmful immune responses to the infection.

nervous system. They all result in nerve injury which causes changes readily detectable on electromyography/nerve conduction testing, in contrast to CNS diseases which do not affect the readout on EMG/NCV testing. All three result in inflammation within the nerves. However, they are also very different. GBS affects almost all nerves to varying degrees, while leprosy and varicella zoster virus (VZV) are much more limited in affecting only a few nerves. The pace of GBS is quite fast, hours to days, while leprosy and VZV are chronic, months to years. And most importantly, GBS is self-limited because the immune system rights itself over time, while in leprosy the damage continues as long as the pathogen remains in the nerve.

4.1 Leprosy

4.1.1 Definition

Leprosy is a chronic infection of predominantly skin, upper respiratory tract, and peripheral nerves caused by the pathogen, *Mycobacterium leprae*. *M. leprae*, the first bacterium to be identified as causing disease in humans, by Gerhard Hansen in 1873, is rod-shaped, and closely related to the bacterium causing tuberculosis, *M. tuberculosis*.

4.1.2 Etiopathogenesis

The neuropathy of leprosy is not well understood, partly because of the lack of good models. The bacterium is present in peripheral nerves; the relative role of the bacterium versus the host response to it in nerve injury is not known and may vary from individual to individual. Patients with leprosy develop both anesthesia, i.e., loss of sensation from nerve damage and pain due to damage to nerves.

The pathogenesis of leprosy depends to a large extent on the magnitude of the host immune response vs. the pathogen, with tuberculoid leprosy on one extreme, lepromatous leprosy on the other extreme, and most cases being somewhere between. In tuberculoid leprosy, there are few organisms but an intense inflammatory response to those few organisms leading to substantial tissue injury. In the lepromatous form, there are large numbers of organisms, but little inflammatory response, and the nerve and skin damage tends to be less. There may be gene expression signatures in infected tissues that differentiate the two ends of this spectrum of disease; genes in the leukocyte immunoglobulin-like receptor (LILR) family were significantly upregulated in lesions of patients with lepromatous forms of leprosy relative to other forms of the disease [26]. Some of the products of these genes are thought to suppress innate host defense mechanisms, possibly by blocking antimicrobial activity triggered by Toll-like receptors (TLRs). Cell-mediated immunity to the bacterium is thought to be the primary means by which bacteria are killed, and patients with low bacteria loads, paucibacillary leprosy, have high levels of anti-bacterial cell-mediated immunity. The reverse appears to be true for anti-bacterial antibody levels which are higher in lepromatous forms with higher bacterial load.

Our understanding of the neuropathy of leprosy has been improved by infecting Schwann cells (SCs) in vitro with *M. leprae*; this cell is the major target cell during active infection. *M. leprae likely* infects nerves by entering through their blood supply and lymphatics; the bacterium binds to and enters SCs. Once inside SCs, the pathogen survives within the cell, does not induce death of the SC, and does not affect the ability in vitro of the SC to associate with axons. In the tuberculoid form of the disease (see Sect. 4.1.4 below), in which the number of bacteria is relatively low, nerve injury occurs more severely and more rapidly, implicating strong cellular immune response as being more important in nerve damage. Nerve involvement is less dramatic in the lepromatous form, in which the number of organisms is very high, but the cellular immune response is weaker.

4.1.3 Pathology

As in the pathology of its cousin, tuberculosis, the hallmark of pathology in leprosy is the granuloma. These lesions are composed of giant cells consisting of fused, multinucleated macrophages, surrounded by epithelioid cells, plasma cells, and fibroblasts. The nerve is frequently enlarged and can be readily palpated. Although the usual involvement is in the peripheral nerve, ganglia such as the dorsal root, sympathetic and trigeminal ganglia can be involved. The macrophage is the reservoir cell for *M. leprae* and can contain as many as 100 *M. leprae* organisms per cell.

4.1.4 Genetics/Epidemiology/Classification

Millions of individuals worldwide develop leprosy and the number of newly infected people per year is about a quarter million. The disease is at least 2,000 years old since *Mycobacterium leprae* DNA was found in a first century CE tomb in Israel. India has the most cases followed by Brazil and Burma. There has been considerable interest in determining genetic risks for leprosy since the organism is present widely in communities but only a small percentage of individuals become chronically infected. It is thought that possibly genetic factors are responsible for clinical differences in patients with leprosy.

4.1.5 Clinical Manifestations

Most patients with leprous neuropathy have a gradual onset of symptoms with either dysesthesias or neuropathic pain. Cooler areas are preferentially involved, such as the dorsum of the forearm, but not the antecubital fossa. Facial and trigeminal involvements are frequently seen. Many nerves can become enlarged and palpable. The ulnar nerve is most frequently involved and damage to nerves in the arm can lead to hand deformities (see Fig. 12.9).

The natural history of the disease is highly variable and is determined by the type of leprosy. Tuberculoid leprosy has early damage to nerves which may be severe, but often the illness can be self-limited even without therapy. In contrast, untreated lepromatous leprosy frequently progresses to deformity, blindness, and at times death.

4.1.6 Diagnosis/Therapy

Diagnosis is usually is made by the confluence of symptoms and signs. Laboratory confirmation, however, is not straightforward. Some centers in endemic areas are using an ELISA for a *M. leprae* antigen called phenolic glycolipid-1 (PGL-1). As in tuberculosis, multidrug therapy (MDT) is now used. Rifampicin, dapsone, and clofazimine taken for 1 year are one recommended combination. Single-dose MDT of rifampicin, ofloxacine, and minocycline can be used a single lesion.

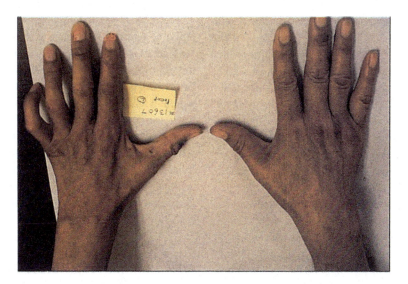

Fig. 12.9 "Clawed hand" in lepromatous leprosy. The fourth and fifth fingers on the left are "clawed" as a result of paralysis of the ulnar nerve the most common trunk involved pathologically in leprosy. The involvement of the ulnar nerve also results in the loss of sensation in these fingers

4.2   VZV Infection

Infection with the herpes virus, VZV, is a relatively common infectious cause of neuropathy in Western civilization [27]. In many respects, VZV neuropathy also called shingles mimics leprosy, the prototypic infectious neuropathy. In both diseases, the pathogen is chronically present in the nerve, inflammation contributes a great deal to clinical disease, and effective treatments are available. The virus, VZV, has some similarities to that of HSV (see Sects. 3.1 through 3.6 in this chapter). Both HSV and VZV have relatively large genomes and are members of the herpesviridae family of double-stranded DNA viruses. The initial infection is usually acquired during childhood as chickenpox, and after clearance of the acute infection, VZV becomes latent in neurons only. Clinically relevant reactivation occurs as a skin rash called herpes zoster, or shingles. The rash occurs in the forms of vesicles occurring unilaterally in a dermatomal distribution as shown in Fig. 12.10. Initially, the vesicles are clear, but then become turbid and crust within 5–10 days. Shingles is sometimes preceded by sensory symptoms such as pain, tingling, or numbness. The pain of the rash can be severe, and sometimes pain or sensory symptoms can occur without a rash, which is called zoster sine herpete. Symptoms usually resolve after resolution of the rash, but sometimes pain can continue or even rarely worsen after rash resolution, which is known as post-herpetic neuralgia. Headache and fever can be a prominent, early manifestation of shingles. The diagnosis is almost always made by clinical evaluation, but laboratory support can be derived from demonstrating high-IgM anti-VZV antibody levels in blood or a positive identification of the virus in fluid from the vesicles either by PCR or by electron microscopy.

VZV can present in a disseminated, aggressive form in adults who did not have chickenpox as a child. Recently, a death was caused by disseminated primary varicella zoster infection in a patient with no history of chicken pox and a negative baseline varicella zoster antibody titer who was in a clinical trial in MS and received a relatively high dose of a new drug for MS, fingoli-

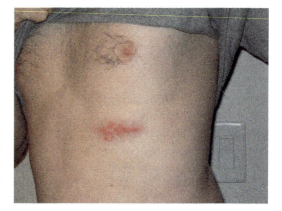

Fig. 12.10 Herpes zoster ("shingles"). This skin eruption is due to reactivation of a latent varicella zoster virus (VZV) infection (from Preston Hunt, http://commons.wikimedia.org/wiki/File:Shingles_on_the_chest.jpg)

mod [28] (see Chap. 19). This patient had been exposed to a child with chicken pox while she was receiving corticosteroids as a treatment for a relapse of multiple sclerosis, in addition to the fingolimod. This case was unusual and represents a rare case of acute serious VZV infection in an adult. Clinical VZV infection in an adult is almost always shingles, reactivation of a decades-old chronic VZV infection. VZV vaccines are increasingly being used, both in children and in adults, and are recommended by the CDC.

Shingles is treated in three ways: by using antiviral medications such as aciclovir, valaciclovir (Valrex), or famciclovir (Famvir), by treating the pain, and by reversing the immunosuppression which contributed to the infection. When the patient is taking immunosuppressive medication, reversing immunosuppression is possible, but sometimes, as in the elderly or in terminal HIV/AIDS this is impossible. Shingles should be treated aggressively, because a substantial percentage of patients develop post-herpetic neuralgia, which can be a debilitating condition requiring large doses of narcotics.

4.3   Immunosuppressed Patients

After initial infection, patients with HIV become progressively more immunosuppressed. This is a slow, gradual process and serious immunosuppression putting the HIV patient at risk for opportunistic infections does not usually occur until about 5–10 years after the initial infection. The following are neurological infections commonly seen in HIV-infected patients once their CD4+ cells have dropped below 200/mm³: toxoplasmosis, PML, cryptococcal meningitis, and neurosyphilis. The first two are almost exclusively seen in immunosuppressed individuals, while the latter two can be seen in immunocompetent individuals, but are present more commonly in the HIV-infected patient.

4.3.1 Cerebral Toxoplasmosis

This infection is only seen in patients with significant immunosuppression [29]. At the height of the AIDS epidemic, it was by far the most common CNS manifestation of HIV infection, despite the fact that CNS infection with this pathogen is not seen in immunocompetent individuals and that 10% of the population is chronically infected with this parasite. A patient with this infection has been described earlier in this chapter. The pathogen is *Toxoplasma gondii*, a parasite commonly present in uncooked food or cat feces. The usual first presentation is a seizure in a young person, although focal findings such as hemiparesis or hemisensory findings can also occur. Diagnosis is usually made by characteristic imaging, usually MRI, which reveals multiple gadolinium-enhancing brain lesions in an immunosuppressed patient. The organism is very sensitive to antibiotics, primarily sulfadiazine and pyrimethamine; the imaging abnormalities may completely resolve. It is one of the few CNS infections for which neurologists will routinely use a therapeutic trial as a routine diagnostic technique rather than obtaining biopsy evidence of the infection. However, treatment is not always permanently effective and infection can recur. The prognosis is usually good and mortality is usually from other opportunistic infections or progressions of AIDS.

4.3.2 Progressive Multifocal Leukoencephalopathy

This infection is only seen in patients with significant immunosuppression and in that way is similar to toxoplasmosis. However, PML is seen occasionally in patients immunosuppressed by medications, such as natalizumab, efalizumab, and rituximab [30], while toxoplasmosis has not been described with those medications. The reasons for this disparity are unknown. PML is described earlier in this chapter in Sect. 3.7.3, as a parenchymal CNS infection contrasting with the prototypic CNS parenchymal infection HSE.

4.3.3 Cryptococcal Meningitis

This infection, which can occur in both immunocompetent and immunosuppressed patients, is described earlier in this chapter as a meningitis resembling the prototypic meningitis of Lyme disease.

4.3.4 Neurosyphilis

Treponema pallidum is a spirochete which causes syphilis in humans; transmission is through sexual intercourse. Much of what we now know about neurosyphilis is based on inadequate information, i.e., case reports and small series, which suffer from a number of biases [31]. Our clinical appreciation of this disease is primarily from the pre-antibiotic and pre-HIV eras, when syphilis was more common and occurred exclusively in immunocompetent individuals.

Syphilis has a long history, likely dating back to Roman times. Descriptions of the disease resurfaced during the European wars of the late fifteenth century when Italian, French, and Spanish soldiers spread the "red plague" or the "great pox" throughout Europe, while Spanish and Portuguese sailors spread the disease to the Philippines, India, China, and America. Among other names, it was called the German Pox, the Polish Illness, the Portuguese disease, the Castilian infection, or the French Disease.

The illness begins with a primary infection occurring within days to weeks after sexual contact, often manifested as a chancre, a sore on the penis. The chancre usually resolves completely and is followed by secondary syphilis which is a skin rash often on the palms, but which can occur anywhere (see Fig. 12.11). Early neurosyphilis within the first few months of infection can involve infection in the CNS without symptoms or a typical basilar meningitis, similar to Lyme

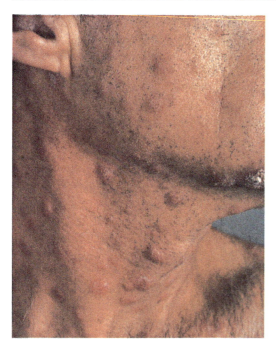

Fig. 12.11 Syphilis. Skin involvement is commonly seen, usually as a relatively early manifestation of the disease. The infection is readily cleared at early stages with antibiotics

meningitis, in which there is headache, malaise, stiff neck, and nonspecific symptoms. Over time persistent inflammation in the subarachnoid space from syphilitic meningitis can involve arteries and meningovascular syphilis can result. In this form of syphilis, stroke is common, although the development of focal lesions tends to be more gradual than in stroke from classical atherosclerotic mechanisms. Rarely, neurosyphilis within the first year can also present with a gumma, a mass lesion frequently occurring in the posterior fossa, appearing as a contrast-enhancing mass on imaging adjacent to the dura and frequently with associated edema.

Late neurosyphilis is not commonly seen now, relative to the pre-antibiotic era. The classical syndromes, seen commonly in the pre-antibiotic era, were general paresis of the insane (GPI) or tabes dorsalis. GPI is a dementing process often associated with significant psychiatric manifestations such as mood changes, grandiose delusions, or marked personality changes. The brainstem is also involved and irregularly shaped pupils that accommodate to near focus, but do not react to

light, also called Argyll Robertson pupils, can be a feature of this stage. The inflammation within the brain is minimal, but meningeal inflammation is present. The most striking symptom in neurosyphilis, lightning pains, is found in tabes dorsalis, another late manifestation of neurosyphilis. These pains are described as bouts of jabbing and lancinating pain usually in the legs. Pains can also occur in the abdomen, leading to a mistaken diagnosis of an acute abdominal surgical emergency. The likely localization is the posterior nerve root, and there is loss of pain, position, and vibration senses and occasionally loss of sensory aspects of bladder. Problems with urination and with balance are common. The CSF in tabes dorsalis and GPI are abnormal with a positive CSF VDRL, elevated protein, and frequently increased number of mononuclear cells.

The diagnosis of neurosyphilis is made by identifying characteristic antibody responses in the CSF and serum. Isolation of the causative spirochete is not routinely attempted because *T. pallidum* is a very difficult bacteria to culture, since it is fastidious and any attempts to culture the pathogen ex vivo have very low yield of positive results. The most sensitive and specific assay for neurosyphilis is a VDRL of the CSF, an inexpensive fast assay which make use of the fact that active infection with *T. pallidum* induces formation of an autoantibody to molecules called phospholipids. VDRL as an antiphospholipid antibody assay and its relationship to the anticardiolipin antibody assay and the antiphospholipid antibody syndrome are discussed at greater length in Chap. 18. A negative FTA/ABS of the serum, an antibody assay which directly tests binding of antibodies to *T. pallidum*, will effectively rule out neurosyphilis, although because of its increased cost, many laboratories will only perform an FTA/ABS assay when the VDRL is positive.

Although neurosyphilis is not restricted to individuals with HIV/AIDS, a large percentage of individuals with neurosyphilis have HIV/AIDS, and in the past few decades neurosyphilis has been increasingly a disease of homosexual men, sometimes labeled as men who have sex with men (MSM). It is unclear whether HIV/AIDS substantially changes the natural history of neurosyphilis. Certainly there is no evidence that CNS invasion occurs more commonly in HIV-positive individuals than those without HIV infection.

The treatment of neurosyphilis consists of penicillin intravenously or intramuscularly. Some clinicians feel that ceftriaxone IV is superior to penicillin but that remains unproven.

References

1. Pachner AR. Lyme disease. Trends Neurosci. 1989;12(5):177–81.
2. Scheld WM, Whitley RJ, Marra CM, editors. Infections of the central nervous system. 3rd ed. Philadelphia: Lippincott-Williams; 2004.
3. Roos KL, editor. Principles of neurologic infectious diseases. New York: McGraw-Hill; 2005.
4. Pachner AR, Dail D, Bai Y, et al. Genotype determines phenotype in experimental Lyme borreliosis. Ann Neurol. 2004;56(3):361–70.
5. Pachner AR, Basta J, Delaney E, Hulinska D. Localization of Borrelia burgdorferi in murine Lyme borreliosis by electron microscopy. Am J Trop Med Hyg. 1995;52(2):128–33.
6. Kaiser R. Intrathecal immune response in patients with neuroborreliosis: specificity of antibodies for neuronal proteins. J Neurol. 1995;242(5):319–25.
7. Clark JR, Carlson RD, Sasaki CT, Pachner AR, Steere AC. Facial paralysis in Lyme disease. Laryngoscope. 1985;95(11):1341–5.
8. Pachner AR, Steere AC. The triad of neurologic manifestations of Lyme disease: meningitis, cranial neuritis, and radiculoneuritis. Neurology. 1985;35(1):47–53.
9. Hansen K, Lebech AM. The clinical and epidemiological profile of Lyme neuroborreliosis in Denmark 1985-1990. A prospective study of 187 patients with Borrelia burgdorferi specific intrathecal antibody production. Brain. 1992;115(Pt 2):399–423.
10. Steere AC, Pachner AR, Malawista SE. Neurologic abnormalities of Lyme disease: successful treatment with high-dose intravenous penicillin. Ann Intern Med. 1983;99(6):767–72.
11. Ljostad U, Skogvoll E, Eikeland R, et al. Oral doxycycline versus intravenous ceftriaxone for European Lyme neuroborreliosis: a multicentre, non-inferiority, double-blind, randomised trial. Lancet Neurol. 2008;7(8):690–5.
12. van de Beek D, de Gans J, Tunkel AR, Wijdicks EF. Community-acquired bacterial meningitis in adults. N Engl J Med. 2006;354(1):44–53.
13. Jarvis JN, Harrison TS. HIV-associated cryptococcal meningitis. AIDS. 2007;21(16):2119–29.
14. Be NA, Kim KS, Bishai WR, Jain SK. Pathogenesis of central nervous system tuberculosis. Curr Mol Med. 2009;9(2):94–9.
15. Anderson NE, Willoughby EW. Chronic meningitis without predisposing illness – a review of 83 cases. Q J Med. 1987;63(240):283–95.
16. Savarin C, Bergmann CC. Neuroimmunology of central nervous system viral infections: the cells, molecules and mechanisms involved. Curr Opin Pharmacol. 2008;8(4):472–9.
17. Pessach I, Walter J, Notarangelo LD. Recent advances in primary immunodeficiencies: identification of novel genetic defects and unanticipated phenotypes. Pediatr Res. 2009;65(5 Pt 2):3R–12.
18. Skoldenberg B, Forsgren M, Alestig K, et al. Acyclovir versus vidarabine in herpes simplex encephalitis. Randomised multicentre study in consecutive Swedish patients. Lancet. 1984;2(8405):707–11.
19. Whitley RJ, Alford CA, Hirsch MS, et al. Vidarabine versus acyclovir therapy in herpes simplex encephalitis. N Engl J Med. 1986;314(3):144–9.
20. Singer EJ, Valdes-Sueiras M, Commins D, Levine A. Neurologic presentations of AIDS. Neurol Clin. 2010;28(1):253–75.
21. Robertson KR, Su Z, Margolis DM, et al. Neurocognitive effects of treatment interruption in stable HIV-positive patients in an observational cohort. Neurology. 2010;74(16):1260–6.
22. Muller M, Wandel S, Colebunders R, Attia S, Furrer H, Egger M. Immune reconstitution inflammatory syndrome in patients starting antiretroviral therapy for HIV infection: a systematic review and meta-analysis. Lancet Infect Dis. 2010;10(4):251–61.
23. Tan K, Roda R, Ostrow L, McArthur J, Nath A. PML-IRIS in patients with HIV infection: clinical manifestations and treatment with steroids. Neurology. 2009;72(17):1458–64.
24. Gyure KA. West Nile Virus Infections. J Neuropathol Exp Neurol. 2009;68(10):1053–60.
25. Lipton HL. Human Vilyuisk encephalitis. Rev Med Virol. 2008;18(5):347–52.
26. Bleharski JR, Li H, Meinken C, et al. Use of genetic profiling in leprosy to discriminate clinical forms of the disease. Science. 2003;301(5639):1527–30.
27. Steiner I, Kennedy PG, Pachner AR. The neurotropic herpes viruses: herpes simplex and varicella-zoster. Lancet Neurol. 2007;6(11):1015–28.
28. Cohen JA, Barkhof F, Comi G, et al. Oral fingolimod or intramuscular interferon for relapsing multiple sclerosis. N Engl J Med. 2010;362(5):402–15.
29. Cohen BA. Neurologic manifestations of toxoplasmosis in AIDS. Semin Neurol. 1999;19(2):201–11.
30. Tan CS, Koralnik IJ. Progressive multifocal leukoencephalopathy and other disorders caused by JC virus: clinical features and pathogenesis. Lancet Neurol. 2010;9(4):425–37.
31. Marra CM. Update on neurosyphilis. Curr Infect Dis Rep. 2009;11(2):127–34.

Both chronic systemic inflammatory diseases and neurological diseases occur frequently so an occurrence of both in a single patient may be due to chance alone. However, there are a number of systemic inflammatory diseases which commonly involve the nervous system. Selected ones which are most likely to be seen by a practicing neurologist in the community are briefly summarized below. A more thorough review of neurological involvement in patients with rheumatic disease was recently published by Sofat et al. [1]. Rheumatic disease, or rheumatism, is a term used to denote the field of medicine of connective tissue and joint disorders, such as rheumatoid arthritis, systemic lupus erythematosus (SLE), vasculitides, and other diseases in which inflammation is prominent and a precise etiology has not been defined.

1 Neurosarcoidosis

Sarcoidosis is a relatively common idiopathic inflammatory disease, which can affect many different systems of the body. There are many similarities of sarcoidosis with tuberculosis, and it seems likely that ultimately a pathogen responsible for sarcoidosis will be identified, possibly one similar to Mycobacterium tuberculosis. Usually, patients have symptoms referable to the involvement of the lung, eye, liver, or skin. Although isolated CNS involvement without systemic sarcoidosis is unusual among all sarcoidosis patients, it is not that unusual for a neurologist to be faced

with this problem, since sarcoidosis is so common. The life-time risk for sarcoidosis in African-Americans, the most commonly involved group, is as high as 1 in 40 individuals. Neurosarcoidosis as a mimic of multiple sclerosis has already been described in Chap. 6.

The most common location within the CNS affected by inflammation and granuloma is the meninges (see discussion on the meninges in Chap. 2), although the granulomas can occur anywhere in the CNS. Involvement of the cranial nerves and chronic meningitis of the base of the brain are common, as is optic nerve involvement (see Fig. 13.1) and the MRI of the brain may reveal areas of demyelination or gadolinium enhancement. Risk/benefit assessment of treatment with immunosuppressives must be carefully performed, since inflammation and symptoms can resolve spontaneously and "adverse effects associated with high-dose systemic corticosteroids, the standard therapy, have discouraged practitioners from initiating treatment in the absence of significant symptomatic neurologic disease" [2].

2 Systemic Lupus Erythematosus

SLE, a relatively common autoimmune disease affecting multiple organ systems, frequently involves the nervous system, and neurological complications have been reviewed extensively by both rheumatologists [3] and neurologists [4]. In the latter review, 128 unselected patients with

A.R. Pachner, *A Primer of Neuroimmunological Disease*,
DOI 10.1007/978-1-4614-2188-7_13, © Springer Science+Business Media, LLC 2012

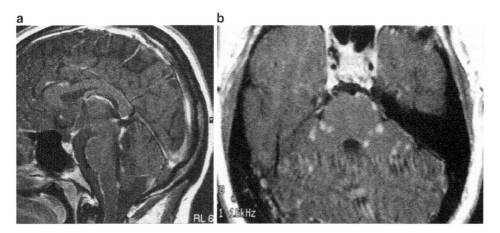

Fig. 13.1 Neurosarcoidosis. (**a**) Sagittal and (**b**) axial post-gadolinium T1W images demonstrating thick, irregular, and nodular leptomeningeal enhancement, which is more striking along the basilar and perimesencephalic cisterns, affecting the cisternal segments of the cranial nerves. There are also enhancing nodules along the courses of the cranial nerves

SLE, 94% female, were evaluated neurologically, and one or more neuropsychiatric syndromes were present in 102 (80%). The most common was headache in 57%, moderate to severe cognitive impairment in 36%, and sensorimotor polyneuropathy in 22% but psychiatric problems were also common, especially severe anxiety or depression. Although many rheumatologists had hoped some of the new recombinant monoclonal antibodies available for targeting specific parts of the immune system (see Chap. 19) would have efficacy in SLE, recent clinical studies in SLE have been disappointing and there have been no major breakthroughs, indicating that neurological manifestations of SLE will continue to be a major problem into the foreseeable future.

3   Antiphospholipid Antibody Syndrome

Some individuals with SLE or other autoimmune or inflammatory processes are found to have autoantibodies to phospholipids in their blood, sometimes called secondary antiphospholipid antibody syndrome (APS) (see Inset 13.1). Others develop antiphospholipid antibodies without a clear antecedent or causative process, called primary APS. Research criteria for APS have been developed [5], but many clinicians entertain the diagnosis when a single test, either anti-cardiolipin antibody or lupus anticoagulant, is positive in patients with stroke or venous thrombosis. Antiphospholipid antibodies are usually detected either as anti-cardiolipin antibodies in a standard ELISA (see Chap. 18 for a description of ELISAs) or in an assay known as lupus anticoagulant (LA). In the LA test, clotting of the blood in the dilute Russell's viper venom time assay (dRVVT) (Fig. 13.2) is delayed by the presence of a serum factor. Other tests which are frequently positive in APS are the autoantibody assays called the Venereal Disease Research Laboratory (VDRL) test or the Rapid Plasma Reagin (RPR) tests commonly used for screening for syphilis. These tests are positive in syphilis because autoantibodies against phospholipids are generated during active syphilis infection. Anti-cardiolipin antibodies that develop during syphilis infection can bind to cardiolipin independent of the presence of β-2 glycoprotein, a serum protein. In contrast, anti-cardiolipin antibodies that develop in APS require this serum factor for binding. How anti-cardiolipin antibodies are triggered in APS and whether this process bears any resemblance to anti-cardiolipin development in syphilis are unknown.

Fig. 13.2 Russell's viper—the snake providing the venom for the dilute Russell's viper venom time assay. Dilute venom from this viper results in clotting of blood from normal humans in under 30 s, but the time is prolonged when a lupus anticoagulant is present. (Figure is by Stefan STS, in public domain at http://commons.wikimedia.org/wiki/File:20090613_Daboia_russelii_Russells_Viper_Kettenviper_Pune.jpg)

APS can be a cause of venous hypercoagulability and theoretically could predispose to cerebral venous sinus thrombosis, but whether APS is linked to neurological disease is controversial. For many years, it was thought that APS predisposed to routine thrombotic arterial cerebrovascular disease, such as atherosclerotic stroke syndromes, but recent data makes this appear less likely. In 1,700 patients included in the Antiphospholipid Antibodies and Stroke Study (APASS), a prospective cohort study within the Warfarin vs. Aspirin Recurrent Stroke Study (WARSS), the presence of antiphospholipid antibodies was not predictive of recurrent thrombotic events [6]. The investigators in the WARSS study felt that routine screening for antiphospholipid antibodies in patients with ischemic stroke did not appear warranted. However, the investigators in this study were still concerned enough about the presence of these antibodies that they recommended that patients with first ischemic stroke and a single positive antiphospholipid antibody test result be anticoagulated with either aspirin or moderate doses of warfarin.

4 Sjogren's Syndrome

Sjogren's syndrome, first described in 1933 by Henrik Sjogren, a Swedish ophthalmologist, is an idiopathic chronic inflammatory disease, more common in women, characterized by inflammatory infiltrates in salivary and lachrymal glands. These glands normally provide lubricating fluid to the mouth and eyes, and Sjogren's syndrome characteristically causes dry mouth and dry eyes. The syndrome can either be secondary, i.e., associated with other inflammatory diseases, or primary occurring on its own. The inflammatory infiltrates can occur in other tissues such as the nervous system, leading to neurological manifestations in as many as 20–25% of patients with Sjogren's. Most neurologists and rheumatologists accept the occurrence of peripheral nervous system manifestations, primarily peripheral neuropathy, as being a frequent manifestation of the disease. This neuropathy is usually a classical stocking-glove axonal sensorimotor polyneuropathy but can be a pure sensory neuropathy or inflammatory involvement of multiple unpredictably distributed nerves, called mononeuritis multiplex. Cranial nerves can be involved, usually the eighth. Symptomatic muscle involvement can also occur. According to two neurologists, one in the USA [7] and the other in France [8], CNS manifestations of Sjogren's can also occur, although this manifestation of the syndrome is more controversial than the peripheral neuropathy. CNS manifestations have been reported to consist of myelopathies, encephalopathies, seizures, or focal involvement of acute onset mimicking stroke or multiple sclerosis. Oligoclonal bands (OCBs) in the CSF (see Chap. 5 for a description of OCBs) can occur, but are less common than in MS. Anti-SS-A or anti-SS-B antibodies occur in 50% of patients with Sjogren's CNS presentations, but almost never in patients with MS. The diagnosis of Sjogren's syndrome with MS-like disease is important, because it can have a very good response to cyclophosphamide therapy (see Chap. 19 for a description of cyclophosphamide).

5   Vasculitis

The term "vasculitis" is used loosely by many neurologists and other clinicians. The term literally means "blood vessel inflammation," and, as such, can be a secondary feature of many diseases. For instance, varicella zoster infection is associated with at least two forms of vasculitis: a large vessel unifocal granulomatous arteritis and small vessel multifocal vasculopathy. Any chronic meningitis, such as cryptococcal meningitis, can result in inflammation of blood vessels which pass in the subarachnoid space. In contrast to these vasculitides which are secondary to another disease process, vasculitis affecting the nervous system can rarely occur as a primary vasculitis.

Primary vasculitis affecting the nervous system is almost always a nervous system manifestation of a systemic vasculitis [9]. Many systemic vasculitides, which as a group are relatively rare disorders, result in neurological involvement, and can be divided into CNS or PNS involvement. The pattern of PNS involvement is usually a mononeuritis multiplex. This is because the blood vessels to the nerve, i.e., the vasa nervorum, can be involved, which causes ischemia of the nerve supplied by that blood vessel. PNS involvement can also consist of a distal symmetrical sensorimotor polyneuropathy. In contrast, parenchymal CNS blood vessel involvement tends to be quite rare among these rare disorders, but when it does occur, the CNS involvement tends to demonstrate a broader spectrum than the PNS involvement. Diseases within this group include vasculitides affecting large blood vessels (giant-cell arteritis and Takayasu's arteritis), medium-sized blood vessels (polyarteritis nodosa and Kawasaki's disease), and smaller vessels (Wegener's granulomatosis and Churg-Strauss syndrome).

Of these, giant cell arteritis (GCA) is by far the most common, affecting approximately 1 per 10,000 individuals over the age of 50. Symptoms are due to decreased blood flow in affected blood vessels, and include headache, scalp tenderness, fever, and pain in the jaw upon chewing. The most common severe neurological complication is decreased visual acuity from decreased blood flow to the optic nerve. Stroke can also be a complication of GCA. PNS involvement in GCA is uncommon. GCA can overlap with a diffuse muscle aching syndrome in the older population called polymyalgia rheumatica (PMR) (see Chap. 11). The diagnosis of GCA with or without PMR is made by consistent clinical syndrome associated with a highly elevated erythrocyte sedimentation rate (ESR) and anemia. The recommended treatment for GCA is high-dose corticosteroids.

The etiopathogenesis of vasculitides is unknown, although, in some forms, there is increasing evidence that autoantibodies are pathogenic. ANCA, anti-neutrophilic cytoplasmic antibodies, divided into perinuclear and cytoplasmic or p- and c-ANCA on the basis of their staining characteristics, are IgG antibodies which bind to myeloperoxidase (MPO) and proteinase 3, antigens which are normally intracellular. However, there is evidence that these intracellular proteins can be induced by cytokines to appear on the surface where they can be targets of circulating autoantibodies, possibly during inflammation. Anti-MPO antibodies have been demonstrated to be pathogenic in animal models of vasculitis. This research is directly applicable to other inflammatory diseases which may be autoantibody-mediated, especially since it demonstrates that under certain conditions the autoantigen target can be a molecule which under normal conditions is an intracellular molecule, but under unusual conditions can be expressed on the surface, and thus be a target of autoantibodies.

Very rarely, vasculitis can be a primary phenomenon only within the nervous system, and primary nervous system vasculitis can either affect the PNS or CNS. Kissl and colleagues reviewed 350 biopsy-proven cases of PNS vasculitis and were able to identify 7 (2%) in which clinical manifestations were only found in the

PNS and no other tissue [10]. Primary CNS vasculitis, i.e., affecting the CNS without extra-CNS manifestations, is even rarer, although for some reason neurologists find the disease fascinating and frequently make the diagnosis erroneously (see Inset 12.3). Vasculitis only affecting the CNS and no other tissue is called "primary angiitis of the CNS (PACNS)" or sometimes granulomatous vasculitis of the CNS. In PACNS, there is vessel wall inflammation in small and medium vessels of the brain parenchyma, spinal cord, and meninges. The disease usually affects middle-aged otherwise healthy individuals and is subacute with headache, higher cognitive dysfunction, seizure, strokes, and cerebral hemorrhages being in the spectrum. Two excellent reviews have recently been published by American and French physicians [11, 12]. Both articles discuss the fact that PACNS is often confused with other diseases or syndromes, especially reversible cerebral vasoconstriction syndromes, such as those which occur with subarachnoid hemorrhage or meningitides due to chronic infections. Angiography for this disorder has a high false-positive rate, and diagnosis should be made by a combination of clinical criteria, LP and MRI findings. Other more common diseases need to be carefully ruled out for two reasons: PACNS is very rare and its prognosis is usually poor. The recommended treatment is with potent immunosuppressives such as cyclophosphamide. At times, cyclophosphamide can have a dramatic effect on PACNS (see Inset 13.2).

Inset 13.1 Uncontrollable Arm Movements in a 19-Year-Old College Student

---

RQ was a 19-year-old college sophomore who presented with a complaint of uncontrollable movements of her left arm. She was well until 1 year prior to presentation when she had an occasional uncontrolled, involuntary tossing movement of her left arm occurring about once every week. Over the few months prior to presentation, the frequency had increased and she found that she could not study in the library because occasionally her movements would hit someone or something near her and disrupt others. She was initially seen by a psychiatrist and referred to neurology. The remainder of the history was negative; there were no illnesses and no positive family history of movement disorders except for an uncle with Parkinson's disease. She was a college tennis player and did not notice any decreased playing ability over the previous year, although occasionally she would lose a point because her involuntary movements would throw off her game.

Her examination was negative for any abnormalities, but laboratory evaluation revealed a low red and white blood cell count. An antinuclear antibody assay, a blood test frequently positive in SLE, was strongly positive as was the VDRL, a blood test for syphilis. The serum FTA-ABS test was negative, indicating that the positive VDRL was a biological false positive (BFP); see Chap. 18 for an explanation of BFP tests. Further evaluation revealed that she had SLE with secondary APS. MRI imaging of her brain, electroencephalography looking for seizure disorder, and lumbar puncture were normal. The diagnosis was felt to be a small basal gangliar stroke associated with her secondary APS. Treatment with daily corticosteroids resulted in a decreased frequency of her movements, but side effects of the drugs forced withdrawal of this therapy. Eventually, without therapy, the movements decreased over time to less than once a month on last follow-up 1 year after presentation.

Inset 13.2 Headache and Confusion in a 58-Year-Old-Engineer

JG was a 58-year-old engineer initially seen by our University Hospital in February of 2009. He was well and a highly functioning engineer until 11/2007 when he developed a headache. This progressed in severity, and in early 2008 he was seen by a community neurologist who identified by imaging a right temporal lesion, felt to be consistent with a malignancy. A lumbar puncture revealed a CSF under normal pressure, but with a lymphomonocytic pleiocytosis, and a negative cytology. In the late spring of 2008, he underwent a resection of the lesion. The pathology was read as "granulomatous inflammation; focal necrosis with perivascular lymphoplasmacytic infiltrating features suggestive of inflammatory process." Work-up for a systemic process was negative, including CT scan of the chest and blood sedimentation rate were normal. He was treated at first with corticosteroids, but in the early winter of 2008–2009, he had worsening of his brain MRI with involvement of both temporal lobes and his confusion and memory deficit were worse. On examination when I first saw him in February 2009, he was only oriented 1/3, knowing his name only. He did not know where he was, and could not identify the date, month, or year, thinking it was 2014. The president was "Bush," but he identified Obama when prompted with the president's first name. He remembered 0/3 objects at 2 min. The rest of the neurological examination outside of the mental status examination was normal. The MRI of the brain showed a bilateral, extensive temporal lobe process with diffuse, but spotty, contrast enhancement. The history, examination, biopsy, and MRI scan were consistent with PACNS. Cyclophosphamide therapy was instituted immediately, one infusion every 3 weeks,

with dramatic improvement in both clinical and MRI manifestations of disease. As of the summer of 2010, the patient had a nearly normal mental status examination and all MRI lesions had cleared up with the exception of one small area.

References

1. Sofat N, Malik O, Higgens CS. Neurological involvement in patients with rheumatic disease. QJM. 2006; 99(2):69–79.
2. Terushkin V, Stern BJ, Judson MA, et al. Neurosarcoidosis: presentations and management. Neurologist. 2010;16(1):2–15.
3. Feinglass EJ, Arnett FC, Dorsch CA, Zizic TM, Stevens MB. Neuropsychiatric manifestations of systemic lupus erythematosus: diagnosis, clinical spectrum, and relationship to other features of the disease. Medicine (Baltimore). 1976;55(4):323–39.
4. Brey RL, Holliday SL, Saklad AR, et al. Neuropsychiatric syndromes in lupus: prevalence using standardized definitions. Neurology. 2002;58(8): |1214–20.
5. Kaul M, Erkan D, Sammaritano L, Lockshin MD. Assessment of the 2006 revised antiphospholipid syndrome classification criteria. Ann Rheum Dis. 2007;66(7):927–30.
6. Levine SR, Brey RL, Tilley BC, et al. Antiphospholipid antibodies and subsequent thrombo-occlusive events in patients with ischemic stroke. JAMA. 2004; 291(5):576–84.
7. Alexander EL, Provost TT, Stevens MB, Alexander GE. Neurologic complications of primary Sjogren's syndrome. Medicine (Baltimore). 1982;61(4):247–57.
8. Delalande S, de Seze J, Fauchais AL, et al. Neurologic manifestations in primary Sjogren syndrome: a study of 82 patients. Medicine (Baltimore). 2004;83(5): 280–91.
9. Rossi CM, Di Comite G. The clinical spectrum of the neurological involvement in vasculitides. J Neurol Sci. 2009;285(1–2):13–21.
10. Kissel JT, Slivka AP, Warmolts JR, Mendell JR. The clinical spectrum of necrotizing angiopathy of the peripheral nervous system. Ann Neurol. 1985; 18(2):251–7.
11. Birnbaum J, Hellmann DB. Primary angiitis of the central nervous system. Arch Neurol. 2009;66(6): 704–9.
12. Neel A, Pagnoux C. Primary angiitis of the central nervous system. Clin Exp Rheumatol. 2009;27 (1 Suppl 52):S95–107.

The Neuroimmunology of Cancer: Paraneoplastic Syndromes and Primary CNS Lymphoma

14

Cancer represents a major, prolonged immunological disruption similar to that of infection. This chapter does not address the relatively common clinical scenarios of either primary cancers of the brain, which usually are tumors of the glial lineage, or of systemic malignances that metastasize to the brain, such as lung or breast cancers; a discussion of these processes is beyond the scope of this book. Instead, this chapter primarily reviews the much rarer paraneoplastic syndromes, neurological involvement associated with immunological alterations caused by a distant tumor. This chapter also briefly summarizes primary CNS lymphoma since this malignancy can mimic neuroinflammatory disease.

1 Paraneoplastic Neurological Disorders

1.1 Definition

Most neurological problems associated with cancer are caused by metastasis of the cancer to the brain or spinal cord, or compression from cancers pressing on neural tissue, but some are due to "remote" effects of the cancer on the nervous system, conditions usually referred to as "paraneoplastic" syndromes. A section of Chap. 9 which discussed tumors producing monoclonal immunoglobulins causing injury to peripheral nerves is one example of how a cancer, or a premalignant condition, can "remotely" affect the nervous system. In paraneoplastic syndromes affecting the CNS, the abnormal immunoglobulins are usually not associated with monoclonal antibodies, but detected by specialized tests of polyclonal antibody in the serum, referred to as paraneoplastic antibodies. In most paraneoplastic syndromes, autoantibodies are thought to be related to pathogenesis, but the evidence for this in most paraneoplastic CNS syndromes is incomplete.

1.2 Etiopathogenesis

It is likely that most paraneoplastic neurological disorders (PNDs) are immune-mediated. The evidence for this is the presence of detectable autoantibodies to CNS tissue in the serum of these patients at levels clearly greater than in normals or in patients with malignancies who do not have neurological disease. These autoantibodies can occur in patients without demonstrated malignancies, in which case they are also associated with neurological disease. In addition, the autoantibodies bind to neuronal proteins that frequently are expressed by the patient's cancer. Details of these autoantibodies are given in Sect. 1.8 of this chapter and in Chap. 18. These autoantibodies are not clearly pathogenic in that the evidence that they actually cause the pathology is not strong, but their presence is helpful for diagnosis. Antibody assays in PNDs are classified into two large groups. First, antibodies against cell-surface molecules, such as ion or neurotransmitter channels [potassium, calcium, or *n*-methyl-D-aspartic acid (NMDA) receptor], may contribute to pathogenesis. Second, antibodies to intracellular proteins that do not have access to the target antigen are less likely to be pathogenic. However, as was discussed briefly in Chap. 13, in the discussion of anti-myeloperoxidase (MPO) antibodies in vasculitides such as Wegener's granulomatosis, intracellular antigens potentially can be expressed on the outside of a cell during inflammation, so the distinction between these two types of autoantibodies may ultimately for some of the autoantigens and their autoantibodies, if it can be demonstrated that "intracellular" antigens can under certain circumstances be expressed on the cell surface.

1.3   Pathology

The pathology in PNDs is variable. In some cases, there is substantial inflammation in the areas involved. Most PNDs are likely antibody-mediated but not all; there have been some instances of cerebellar degeneration with infiltration by T cells in which neurons appear destroyed by the T cells in neuronophagic nodules. Neuronal degeneration is a common finding as is gliosis, the proliferation of astroglia.

Paraneoplastic cerebellar degeneration (PCD) is characterized by loss of Purkinje cells, a cerebellar cell type described in Chap. 2. These cells can be seen in the normal cerebellum as shown in Fig. 14.1a as very large cells with elliptical cell bodies and 2–3 processes in what is called the Purkinje layer at the interface of the granular layer below and the molecular layer above. In Fig. 14.1b in the cerebellum of a patient with PCD, Purkinje cells have been lost in the disease process, presumably targeted by the anti-Purkinje cell antibodies present in PCD, and there is no Purkinje layer, but the granular and molecular layers are normal.

1.4   Genetics and Epidemiology

As life span increases and the prevalence of cancers increases with an increasing number of

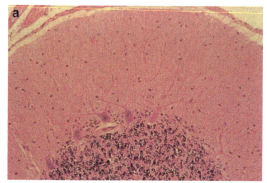

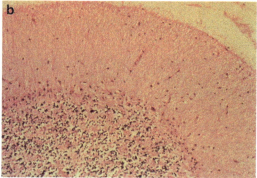

Fig. 14.1 Loss of cerebellar Purkinje cells in PCD. (**a**) Normal cerebellum showing large Purkinje cells. (**b**) Cerebellum from a patient with paraneoplastic cerebellar degeneration showing complete loss of Purkinje cells

elderly in the population, the prevalence of neurological syndromes associated with cancers will continue to increase. In addition, as our tools for detection of these syndromes improve and the awareness of these conditions increases, they will be increasingly recognized. An estimate of their prevalence has been 1 in 10,000 patients with cancer, but this may be too low [1]. There do not appear to be any genetic predispositions, although PNDs are rare enough and sporadic enough that no one has carefully evaluated this issue.

1.5   Clinical Manifestations

These syndromes vary considerably, but the most distinctive are PCD and limbic encephalitis. Lambert–Eaton myasthenic syndrome (LEMS), which was discussed in Chap. 10, is also a characteristic PND.

1.5.1 PCD

The usual first symptom is mild gait unsteadiness which progresses over time, usually weeks to months, so that by the time the patient comes to the evaluation of a neurologist, obvious severe ataxia (see Chap. 2) may be present. Other symptoms of involvement of the cerebellum, such as problems with eye movement, speech, and swallowing, are common. An example of the history of a patient with PCD is in Inset 14.1.

Inset 14.1 68-Year-Old-Woman Who Stumbled

A 68-year-old previously healthy woman began to notice some problems with coordination and occasional stumbling. Over the course of a few weeks, this progressively became worse and she sought medical attention. She saw her primary care physician who could not find any abnormalities and told her to return in 1 month if things did not get better. In 1 month, she was much worse and had had a number of falls. She was referred to a neurologist who found a severe cerebellar ataxia; MRI scan of the brain and lumbar puncture were normal. During her workup, she progressed to the point where she was wheelchair bound. PCD was considered and she underwent evaluation that revealed multiple enlarged lymph nodes and masses in the liver, consistent with metastatic cancer. The liver was biopsied, and a highly malignant cancer was diagnosed pathologically. The primary lesion could not be identified. PCD was confirmed by demonstration of anti-Yo antibodies in the serum. The patient's health deteriorated rapidly due to spread of her malignancy and she died 4 months after the onset of her symptoms. The primary of her metastatic cancer was never identified.

Author's note. This patient had a characteristic presentation of PCD in which the malignancy presented with paraneoplastic neurological disease. Unfortunately, in this patient, the cancer was highly malignant and progressed too rapidly to be treated by the time it was recognized.

1.5.2 Limbic Encephalitis

The hallmark of limbic encephalitis is a combination of affective symptoms, emotional lability or uncharacteristic outbursts, and cognitive problems. Mood alterations, hallucinations, short-term memory loss, and sleep problems can deteriorate to dementia or lethargy. Seizures are not uncommon. An example of a patient with limbic encephalitis associated with anti-NMDA receptor antibodies is presented in Inset 14.2. This syndrome has also been termed "anti-NMDA receptor encephalitis."

Inset 14.2 The Nursing Student Who Became Confused

A 27-year-old nursing student presented to the emergency room with a 3-week history of confusion, aberrant behavior, and seizure-like movements. She was admitted to the neurology department with a diagnosis of status epilepticus (constant seizure activity) with frontal lobe seizures,

She was well until about 6–8 weeks before admission when her schoolwork started deteriorating. Three weeks before admission she began to act in a strange manner. She had previously been quiet and pleasant, but became agitated, belligerent, and talkative. She became confused and disoriented, and then developed strange movements of her mouth and face. There was no headache and no fever. One week before admission friends and family were finally able to get her to the local emergency room, where she was admitted to the psychiatric service with a diagnosis of acute psychotic state versus drug intoxication. She did not respond to antipsychotic medications and in fact worsened, becoming completely disoriented with intermittent episodes of severe lethargy. Her facial movements worsened and she was transferred to the psychiatric service in our hospital, and neurology was asked to consult.

Exam revealed a lethargic, but easily arousable young woman, disoriented to person, place, and time. Frequent mouthing movements of her face were present, but no focal weakness. There were unexplained fluctuations in her pulse rate, blood pressure, and respiratory rate. The history and exam were consistent with a limbic encephalitis. During the next few days, she became unresponsive, but was awake and tracked movement in her room with her eyes. Psychiatrists made a diagnosis of catatonic schizophrenia. The diagnosis of limbic encephalitis was confirmed by the MRI of the brain and spinal tap. MRI of the brain showed increased signal in the medial aspect of both temporal lobes, most clearly evident on the fluid-attenuated inversion recovery (FLAIR) sequences. Lumbar puncture revealed a moderately increased number of white cells, but normal protein and a negative herpes simplex virus PCR. Paraneoplastic antibody panel was sent and CT scanning of her chest and abdomen were normal. CT scanning of her pelvis only showed a slight enlarged left ovary. When the results of paraneoplastic antibody screen came back with positive test for antibodies to the NMDA receptor, the left ovary was biopsied. A teratoma was found and removed. Treatment with corticosteroids and intravenous immunoglobulin produced no clear improvement and she was transferred to a nursing facility.

Author's note. This patient had a limbic encephalitis with some distinctive features. Although most encephalitides are caused by viral infections, limbic encephalitis is commonly caused by a paraneoplastic process. This patient's initial problem was difficulty with her schoolwork, likely because of the short-term memory deficits commonly present. Subsequent behavioral changes, with confusion and agitation as prominent features, are typical of limbic involvement. Also consistent was the autonomic instability.

The lack of fever and seizures and the negative herpes simplex virus made viral etiologies less likely. The catatonia and mouthing movements in the context of the other features was suggestive of a paraneoplastic syndrome associated with antibodies to NMDA receptors and ovarian teratomas [5], which proved on evaluation to be the correct diagnosis. These forms of limbic encephalitis can be responsive to removal of the tumor and immunosuppression; however, these did not seem to benefit our patient.

1.5.3 Dermatomyositis

Dermatomyositis has been described as a paraneoplastic syndrome. The clinical presentation of this muscle inflammation is not different from that of dermatomyositis without a malignancy described in Chap. 11. Patients who develop dermatomyositis, especially if they are older, have an increased incidence of malignancies when compared with the general population, the most common malignancies being breast, lung, pancreas, and colon. Most studies demonstrate an association with malignancy in about 15–25% of cases of dermatomyositis. Factors in dermatomyositis more commonly seen in those with malignancies are older age (greater than 50), rapid onset of skin or muscular symptoms, and skin necrosis.

1.5.4 Opsoclonus

Opsoclonus is an intermittent involuntary conjugate eye movements in all directions, frequently continuous, and usually of low amplitude. Some neurologists call this as "dancing eyes." When it occurs as a PND, it can be associated with quick involuntary movements of the arms or legs, which are sometimes called "dancing feet." When these occur together, they are called opsoclonus-myoclonus or dancing eyes–dancing feet.

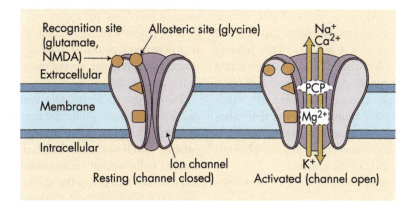

Fig. 14.2 The NMDA receptor. NMDA receptors form a family of receptors that are related in that they are able to bind NMDA and have a heterotetrameric structure. There are four receptor subunits made up of two different types of molecules, NR1 and NR2. Each complete receptor molecule contains at least two NR1 subunits called NR1 and two NR2 subunits; NR2 molecules are further subclassified into NR2A, NR2B, NR2C, or NR2D. These subtypes produce distinct properties, with NR2A and NR2B subunits producing high conductance states, while those with NR2C and NR2D are low conductance. Neurotransmitters, such as γ-aminobutyric acid (GABA); these are large molecules with multiple binding and modulatory sites. NMDARs respond to a host of ligands besides NMDA including glycine, PCP, and Mg^{2+}

1.6 Natural History and Prognosis of PND

The natural history and prognosis are usually completely related to the underlying malignancy. Some PNDs can be completely cured by surgical removal of the underlying malignancy, such as limbic encephalitis associated with ovarian teratoma and anti-NMDA antibodies (see below). Others may linger or progress even after successful excision or other treatment of the tumor.

1.7 Classification

As in many neurological diseases in which etiopathogenesis is mysterious, there are multiple classification schemes for paraneoplastic diseases. Two forms of classification are commonly used. The first is by clinical syndrome, e.g., cerebellar degeneration or limbic encephalitis; these are described above under "Clinical manifestations." The second classification is by presumed pathogenesis.

When paraneoplastic diseases are classified by presumed pathogenesis, one category is antibody-mediated paraneoplastic syndromes, in which antibodies detectable in the serum of these patients react with cell-surface antigens. The most well characterized of these are syndromes of the peripheral nerves or neuromuscular junction, such as LEMS (described in Chap. 10). In contrast, when the autoantigens are intracellular, the pathogenesis is thought to be unknown, or, in some cases, mediated by autoimmune T cells rather than antibodies.

A few examples of diseases in which antibodies against cell-surface antigens are involved are antibodies to:

1. Voltage-gated calcium channels (VGCCs) in LEMS associated with small cell lung cancers.
2. Voltage-gated potassium channels (VGKCs) in limbic encephalitis.
3. NMDA receptors associated with ovarian teratomas in limbic encephalitis (see Fig. 14.2). Despite the fact that the precise mechanisms are not well characterized, it appears likely that these antibodies contribute to pathogenesis. More support for this hypothesis has come from work by Hughes et al. [2], who demonstrated that in anti-NMDA receptor antibody encephalitis, anti-NMDA receptor (NMDAR) antibodies bind to NMDARs in tissue culture

and caused a decrease in NMDAR surface density via a mechanism called capping and internalization (see Chap. 3). This binding selectively decreased NMDAR function in vitro without affecting the function of unrelated channels. These effects were also reproduced by passive transfer of these antibodies into Lewis rats. Interestingly, these antibodies did not affect cell survival or any morphological characteristic of the neurons.

Some paraneoplastic neurological syndromes thought NOT to be due to autoantibodies are opsoclonus (or opsoclonus-myoclonus) associated with neuroblastomas usually, and dermatomyositis associated with a range of malignancies. The mechanisms for these paraneoplastic syndromes are even less well understood than those for the syndromes associated with autoantibodies.

Some of the above syndromes that ARE related to autoantibodies may not always be paraneoplastic. For instance, LEMS patients frequently do not have associated malignancies. Also, Dr. Angela Vincent and colleagues at Oxford have stressed that the encephalitides associated with anti-VGKC antibodies are first, frequently NOT associated with tumors, and second, are highly responsive to immunosuppressive therapies such as plasma exchange and corticosteroids [3].

1.8   Diagnosis

The diagnosis of paraneoplastic syndromes is often obvious when patients with known malignancies present with neurological syndromes. However, not infrequently, patients without histories of cancer present with the neurological symptoms prior to identification of the cancer. In such situations, early diagnosis of a paraneoplastic syndrome can be lifesaving, because excision of the cancer can be curative. However, it frequently requires an astute clinician to consider the diagnosis since paraneoplastic syndromes are rare.

The first steps in the diagnosis, as in other neurological diseases, are a complete history and competent physical examination. Thus, in a middle-aged or elderly individual with no other

medical problems and symptoms of unexplained difficulty with walking in combination with findings of ataxia localizable to the cerebellum, paraneoplastic cerebellar ataxia must be considered in the differential.

Frequently, paraneoplastic syndromes do not cause clear lesions within the nervous system on imaging. However, MRI scans of the brain in patients with limbic encephalitis can show abnormalities in the medial portion of the temporal lobe and patients with cerebellar ataxia may occasionally show some cerebellar atrophy. Many times patients with paraneoplastic syndromes have abnormal CSFs with elevated protein or number of white cells or positive oligoclonal bands.

There are distinctive antibody reactivities in some patients with paraneoplastic syndromes which aid in diagnosis when they are present (see Sect. on "neuronal antigens in paraneoplastic neurological disorders (PNDs)" in Chap. 18). These were originally detected using assays of reactivity of patient sera to whole brain slices, but now many of the target antigens have been purified and are available in recombinant form to allow faster and more accurate diagnostic assays.

1.9   Therapy

Therapy has two major goals: removing the malignancy and decreasing inflammation. Sometimes removal of the malignancy can be curative both of the malignant process and the paraneoplastic syndrome. With respect to decreasing inflammation, many different approaches have been used, but none has been universally successful. Immunosuppressive and immunomodulatory agents commonly used in neuroimmunological diseases are summarized in Chap. 19; many of them have been used in paraneoplastic syndromes.

2   CNS Lymphoma

CNS lymphoma is a malignancy of the brain composed of malignant lymphocytes, usually B lymphocytes. Both the clinical and pathological

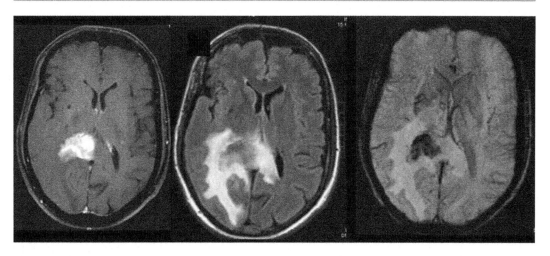

Fig. 14.3 MRI of the brain in primary central nervous system (CNS) lymphoma showing a large single lesion mostly in the right brain but spreading via the corpus callosum to the left

picture can mimic a number of neuroimmunological diseases such as MS or infections. The involvement of the CNS can be either secondary to a lymphoma elsewhere or can be primary (originating in the CNS). The MRI usually demonstrates one or more lesions that enhance with gadolinium, although if the lymphoma presents in the context of AIDS, the lesions may not enhance. Periventricular lesions are not unusual and can mimic MS. However, the gadolinium enhancement is usually diffuse rather than the patchy enhancement frequently seen in MS. Because lymphomas often respond temporarily to corticosteroid therapy, they are frequently confused with other inflammatory processes. The diagnosis of primary CNS lymphoma can sometimes be challenging. The cancer, like other infiltrative brain tumors, can spread along white matter tracts. It usually involves the cerebrum and only rarely the brainstem and spinal cord. Multiple lesions are seen at initial presentation approximately one-third of the time; the appearance of the tumor can be quite dramatic (see Fig. 14.3). The cell type in primary CNS lymphoma is also almost always malignant non-Hodgkin lymphoma, the B-cell type; most of the time pathology will also show an infiltrate of nonmalignant T cells. Treatment is usually with high-dose methotrexate with or without rituximab, although new treatments are being tested [4]. Partial surgical resection has no role and actually decreases survival. Increasing age is a negative prognostic factor.

References

1. Dalmau J, Rosenfeld MR. Paraneoplastic syndromes of the CNS. Lancet Neurol. 2008;7(4):327–40.
2. Hughes EG, Peng X, Gleichman AJ, et al. Cellular and synaptic mechanisms of anti-NMDA receptor encephalitis. J Neurosci. 2010;30(17):5866–75.
3. Vincent A, Buckley C, Schott JM, et al. Potassium channel antibody-associated encephalopathy: a potentially immunotherapy-responsive form of limbic encephalitis. Brain. 2004;127(Pt 3):701–12.
4. Hochberg FH, Baehring JM, Hochberg EP. Primary CNS lymphoma. Nat Clin Pract Neurol. 2007;3(1): 24–35.
5. Dalmau J, Gleichman AJ, Hughes EG, et al. Anti-NMDA-receptor encephalitis: case series and analysis of the effects of antibodies. Lancet Neurol. 2008; 7(12):1091–8.

Neuroimmunology of Degenerative Diseases and Stroke

15

1 Introduction

Degenerative diseases of the nervous system, the most common of which are Alzheimer's disease (AD) and Parkinson's disease (PD), take a huge toll on the older population, as does atherosclerotic cerebrovascular disease (ASCVD). Any potential preventive or therapeutic approach that might have a positive benefit/risk ratio should be pursued to alleviate this burden. There is evidence that these processes have inflammatory components and that down-regulation of key inflammatory components might ameliorate the degenerative process. This research is still in the developmental stage as of 2011 and there are no anti-inflammatory or immunosuppressive medications now being used as a standard of care for any of these diseases, but this may change in the future. The immunology of these diseases is a large and complex area, so only a very brief outline is presented here. Other chronic neurological diseases which may have a immune/inflammatory contribution, such as narcolepsy and amyotrophic lateral sclerosis will not be discussed, since the available neuroimmunological data is even more fragmentary in these diseases.

2 Neuroimmunology of Alzheimer's Disease

AD was first described by Alois Alzheimer (1864–1915), a German psychiatrist and neuropathologist. In 1901, he began studying a 51-year-old woman with severe, progressive memory loss, who died in 1906; Alzheimer obtained her brain and studied its pathology. Using silver stains developed by Franz Nissl and others, Alzheimer correlated his patients' clinical condition, which he called presenile dementia, with its pathological hallmarks. Alzheimer's disease is a now known to be a common, idiopathic disease of the older population which results in progressive dementia. Dementia is literally "loss of mind," the loss of cognitive function. Ability to learn new information is the earliest function lost. Frequently, the disease is first noted by close relatives since patients with AD tend to be apathetic toward their condition. Increasingly, their personalities change and they become more forgetful, confused, and less able to care for themselves. The disease is increasingly common with increasing age, and as many as 1/5 of the population aged 65–74 has AD. The disease must be distinguished from other causes of dementia, such as multiple strokes, also a common cause of dementia in the elderly.

The pathology is limited to the gray matter of brain; other organs are not affected. Neurons are lost, and the increasing neuronal loss results in increasing brain atrophy. The pathological hallmarks on microscopy are senile plaques and neurofibrillary tangles scattered throughout the cerebrum. The senile plaques, also called amyloid plaques, are composed of β-amyloid, abbreviated Aβ, a proteolytic peptide cleaved from a normal membrane protein, amyloid precursor protein (APP). Neurofibrillary tangles are composed of aberrant microtubule-associated proteins, consisting of an excessively phosphorylated

A.R. Pachner, *A Primer of Neuroimmunological Disease*,
DOI 10.1007/978-1-4614-2188-7_15, © Springer Science+Business Media, LLC 2012

normally occurring protein called tau. Many theories on the pathogenesis of Alzheimer's disease exist, ranging from chronic viral infection to acetylcholine dysfunction, but most theories revolve around the excessive plaques and neurofibrillary tangles in the disease.

The immunology of AD has mostly revolved around two issues: inflammatory mechanisms contributing to pathogenesis and the use of vaccines in treatment. There has been some epidemiological evidence that the use of nonsteroidal anti-inflammatory drugs (NSAIDs) provides a protective effect for older individuals from developing AD. This is controversial and is not accepted by all investigators. However, there was adequate interest in this finding to result in controlled studies on treatment of various stages of AD with NSAIDs none of which showed any efficacy. Many still believe, however, that NSAIDs do indeed have an effect but that the drugs need to be used prophylactically prior to disease onset, not as treatments once AD has begun.

One arm of the immune response that appears to be locally activated in plaques is the complement system (see Sect. 2.2.6 of Chap. 1). Senile plaques in AD contain complement proteins. Complement is found early in amyloid deposition, and complement activation is correlated with progression of the disease. AD brains have activated microglia, reactive astrocytes, and increased production of proinflammatory cytokines, all of which support the role of the immune system in the disease. However, increasingly, researchers are also considering the possibility that the immune activation may be neuroprotective rather than harmful. Thus, the jury is still out on whether the immune activation seen in Alzheimer's disease represents a legitimate target for therapy.

Because of the possibility, the amyloid plaques are toxic to the brain and their clearance might ameliorate disease, Elan Pharmaceuticals developed a synthetic Aβ referring to it as AN-1792 and immunized AD patients with this "Alzheimer's vaccine." The study was prematurely halted because of the development of meningoencephalitis in some of the study participants. Although some evidence pointed to less amyloid plaque in immunized patients, there did not appear to be significant clinical benefit with respect to improved cognition or survival. Despite these findings, there is continued research along these lines with the development of a humanized anti-Aβ monoclonal antibody, bapineuzumab, which is being tested in mild and moderate AD.

3 Neuroimmunology of Parkinson's Disease

Parkinson's disease (PD) is nearly as common as Alzheimer's disease in the older population. It is a movement disorder which means that movement is impaired despite normal strength and sensation. The most obvious characteristic is tremor, but the symptom most troubling to the patient is bradykinesia, excessively slow movement because of an inability to effect a smooth rapid movement. Patients say that when their disease is active they feel like they are wrapped in a thin plastic sheet; movement is possible but is slow and labored. Late in the disease, posture is difficult to maintain and cognition can become impaired. As in AD, PD may be a problem with abnormal or excessive proteins in the CNS, in that there is aggregation and accumulation of the protein α-synuclein. A recent hypothesis which has attracted great interest is that α-synuclein stimulates microglial activation which further injures neurons. Microglial activation is a prominent feature of PD, and these cells express HLA-DR molecules and CD11b, which is a component of an integrin molecule (for description of HLA-DR and integrins, see Chap. 1).

As in the instance of amyloid-β in AD, it has not yet been determined definitively whether α-synuclein accumulation is a cause or an effect of the pathology of the disease. The role of synucleins in normal cellular function is unknown. The most dramatic neuropathological finding in PD is loss of neurons, particularly dopaminergic neurons utilizing the neurotransmitter dopamine in the substantia nigra, which is located in the midbrain. Lewy bodies (Fig. 15.1) consist of accumulations of α-synuclein and are characteristic of PD. They are also seen in diseases of the "Parkinson-plus complex," which is Parkinson's

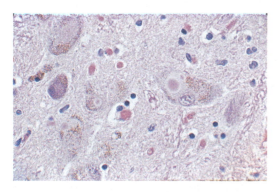

Fig. 15.1 Lewy bodies in the substantia nigra (hematoxylin–eosin stain, magnification ×40). The major pathologic abnormalities in Parkinson's disease (PD) are neuronal cell loss, gliosis, and loss of pigment in the substantia nigra, especially in the ventrolateral portion projecting to the putamen, and the presence of abnormal intracytoplasmic neuronal inclusions called Lewy bodies. Lewy bodies consist of an amorphous central core with a halo of radially arranged neurofilaments measuring 10–20 nm in diameter

disease plus a number of other diseases with which Parkinson's overlaps, such as dementia with Lewy bodies, multiple system atrophy, and progressive supranuclear palsy.

Animal models of Parkinson's disease have provided researchers with opportunities to study its immunology more systematically. The chemical MPTP induces degeneration of the substantia nigra degeneration and clinical disease in nonhuman primates and can induce neuronal dropout in mice, and has been used for many years since its discovery in 1982 by Dr. J.W. Langston [1] (see Inset 15.1). More recently, rotenone, a pesticide, has also been shown to induce Parkinson's disease-like picture in rats. There is some evidence that cellular immunity can exacerbate the disease in these models. In addition, some investigators have found that some inflammatory mediators can affect the progression of cell death.

4   Neuroimmunology of Stroke

Stroke is the sudden death of CNS tissue by a block in blood flow, usually as a result of changes in cerebral blood vessels caused by atherosclerosis. Acute cessation of blood flow in an affected artery can be caused in two ways. The first, and likely most common, is an atherosclerotic embolus formed by rupture of an atherosclerotic plaque, with fragments released from the plaque flowing downstream and lodging in a smaller branch artery causing loss of blood flow. The second mechanism of stroke occurs at the site of the plaque, when the core of the plaque, the atheroma, erodes through the endothelium, and induces a thrombotic occlusion. There is within an atherosclerotic plaque a core containing lipids and debris from dead cells. Surrounding it is a fibrous cap with smooth muscle cells and collagen fibers. Macrophages, T cells, and mast cells populate the plaque and are frequently in an activated state.

Increasingly, investigators of atherosclerosis, in general, and stroke, in particular, consider atherosclerosis as an inflammatory disorder of the blood vessel. The initial insult to the blood vessel wall is believed to be the accumulation of lipid-laden macrophages, referred to as foam cells, beneath the endothelial lining. Collections of these foam cells are responsible for slightly raised fatty streaks. Over time, extracellular lipid rich in cholesterol, T cells, additional macrophages, and smooth muscle cells accumulate. Dendritic cells, immune cells related to macrophages, have also been shown to be present. The smooth muscle cells produce a collagenous cap over the central atheromatous core. Hemorrhage into the plaque may also occur and contribute to restricted blood flow. A variety of cytokines and adhesion molecules (see Sects. 2.2.3 and 2.2.5 in Chap. 1) are produced by the cells in the plaque and likely contribute to the accumulation of the various inflammatory cells within the plaques. At some point in its evolution, the mature plaque, consisting of a lipid-rich center with its live, dying, and dead cells, becomes unstable and prone to rupture, releasing embolic fragments or locally active prothrombotic components.

Mouse models of atherosclerosis mimic some, but not all, aspects of human atherosclerosis; the most studied of these is the apo-E knockout. Knockout of additional genes encoding certain inflammatory molecules reduce atherosclerosis. Such genes include those encoding chemokines and their receptors, adhesion molecules, cytokine signaling molecules, and T cells, so that both human studies and animal models implicate inflammation as a key element of atherosclerosis (two very good review articles have been recently published [2, 3]).

Two important diagnostic techniques are based on the concept of atherosclerosis as an inflammatory-mediated disease. Positron emission tomography (PET) scanning of patients with atherosclerosis identifies macrophages in the plaque that take up ^{18}F-fluorodeoxyglucose (FDG) and have been shown to be prognostic for an occlusive event. A sensitive assay for serum C-reactive protein (hs-CRP) is increasingly used as an indicator of atherosclerotic inflammation, and the American Heart Association and US Centers for Disease Control and Prevention have defined risk groups for atherosclerosis according to their levels of hs-CRP. Whether CRP is simply an indicator of inflammation or is an inflammatory stimulus in atherosclerosis is a subject of active investigation. Statins widely used to reduce atherosclerotic morbidity and originally introduced to lower serum cholesterol have been shown to have the additional therapeutic property of reducing inflammation. Some clinicians also recommend NSAIDs such as aspirin or ibuprofen to lower atherosclerosis risk. Other investigators have even suggested the use of more aggressive measures, such as IVIg or methotrexate, immunomodulatory therapies described in Chap. 19.

Inset 15.1 The Story of MPTP

Illicit laboratories in the 1970s and 1980s attempted to produce congeners of heroin. One of the congeners, MPPP, was quite successful in inducing the highs of heroin, but a contaminant of the synthesis, MPTP, produced a Parkinson's disease-like syndrome in four drug abusers as described by Dr. J.W. Langston [1]. MPTP is now used to induce experimental models of Parkinsonism.

References

1. Langston JW, Ballard P, Tetrud JW, Irwin I. Chronic Parkinsonism in humans due to a product of meperidine-analog synthesis. Science. 1983;219(4587):979–80.
2. Hansson GK, Libby P. The immune response in atherosclerosis: a double-edged sword. Nat Rev Immunol. 2006;6(7):508–19.
3. McColl BW, Allan SM, Rothwell NJ. Systemic infection, inflammation and acute ischemic stroke. Neuroscience. 2009;158(3):1049–61.

16

Important Rare Neuroimmunological Diseases

All of the following syndromes are rare and have annual incidences of less than 1/1,000,000, but have importance over and above their frequency in the population. These disorders are sometimes classified as "zebras," disorders so unusual that they are normally only considered long down the list of likely suspects. As noted in sect. 6.3 in Chap. 6, the term zebra has its origin in the phrase: "When you hear hoofbeats, think horses, not zebras," i.e., when evaluating a patient with an unknown diagnosis, a physician should remember that common diseases occur commonly and that it is more likely to get an unusual manifestation of a common disease than to have a common presentation of a zebra. However, all neuroimmunologists have to think of zebras occasionally when the diagnostic situation does not bring to mind any "horses." The conditions described below constitute far from an exhaustive list. I encourage neurologists who read this chapter to contact me (pachner@umdnj.edu) to describe their most memorable zebra. These syndromes have been divided into three different categories: infectious/postinfectious, autoimmune, and idiopathic, and provided a short description. At the end of each disease description is a short section on why the disease is important out of proportion to its incidence.

1 Infectious/Postinfectious

1.1 Subacute Sclerosing Panencephalitis

This is a subacute encephalitis usually seen in children or adolescents caused by infection with measles virus (Fig. 16.1). It is very rare in countries in which measles vaccination is the norm, but has an annual incidence in other countries, such as Papua New Guinea and India, as high as 56 per million population. Most investigators feel it is caused by a mutated measles virus that is unable to produce multiple measles proteins, but maintains a limited persistent infection in the CNS.

Initially, SSPE affects the posterior part of the brain, spreads to anterior sections and then affects subcortical areas, brain stem, and spinal cord. Inflammation, demyelination, gliosis, and viral inclusion bodies are hallmarks on pathology. Much of the inflammation is perivascular. Clinically, patients with SSPE develop subacutely over weeks to months intellectual deterioration, decline in school performance, and behavioral abnormalities with associated movement disorders, seizures, and visual abnormalities. Myoclonic jerks, the spasmodic, brief movements of a muscle group are common and

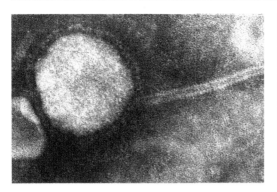

Fig. 16.1 An electron microscope image of measles virus, the cause of SSPE. This virus, within the paramyxovirus family, is quite large for viruses being about 200 nm in diameter. Another group of viruses within the paramyxovirus, distemper viruses, cause a CNS disease in some mammals but not humans, characterized by inflammation, demyelination, and progressive injury resembling multiple sclerosis

Fig. 16.2 George Hoyt Whipple (1878), an American pathologist, researcher, and medical school administrator who was known more for his other achievements than for his description of the disease that bears his name. His research dealt with anemia and the liver; he demonstrated in dogs that liver fed to anemic dogs reversed the anemia. This led to the successful treatment of pernicious anemia in humans by George Minot and William Murphy, who shared the Nobel Prize with Whipple in 1934. He also was a renowned teacher and was the dean of University of Rochester's Medical School for 33 years. In 1907, he described a disease which he called lipodystrophia intestinalis, but which has come to be known as Whipple's disease, which can involve the nervous system (image from Edward G. Miner Library, University of Rochester Medical Center, by permission of Christopher Hoolihan)

are included in diagnostic criteria [1]. Diagnosis is made by recognition of the characteristic clinical picture and confirmed by brain wave tests (electroencephalography or EEG), which shows typical periodic high-voltage synchronous diphasic waves, as well as by high levels of CSF immunoglobulin levels enriched for anti-measles antibody. Prognosis is poor, because of the absence of an effective therapy, and most patients die of the disease. A related disease, which is even rarer and more deadly than SSPE, is measles inclusion body encephalitis (MIBE), which is seen in immunocompromised individuals and usually causes seizures followed by rapidly progressive neurological deterioration.

Importance. SSPE is an example of how a virus infection can cause a chronic inflammatory demyelinating disease. It is also demonstrates how a virus can continue to cause progressive injury despite a strong immune response against the virus., i.e., in SSPE high levels of anti-viral antibody are present both in the blood and within the brain produced by plasma cells within the brain [2].

1.2   Whipple's

Whipple's disease is the ultimate "zebra" (see Fig. 16.2). It is caused by the bacterium, *Tropheryma whipplei*, which is widespread in the environment. The disease usually presents with gastrointestinal symptoms such as abdominal pain, diarrhea, weight loss, malabsorption, and wasting, but it can initially present very rarely as a neurological disease [3]. Approximately one-third of patients with Whipple's disease have neurological involvement at some point in their disease. Neurologically involved patients frequently have encephalopathies, but a number of presentations in the CNS are possible. Lesions in the CNS which enhance with gadolinium contrast are sometimes seen in the brain, and about half of neurologically involved patients have a lymphocytic pleocytosis in the CSF. The organism cannot be cultured, so it must be identified with PCR, usually of biopsy tissue or sometimes of CSF, or with immunohistochemistry of affected tissue or of the duodenum. Some patients are diagnosed based on periodic acid-Schiff (PAS)-positive inclusion bodies, in the absence of any other cause of their illness, in addition to a positive therapeutic trial with antibiotics. Treating Whipple's disease with the antibiotic trimethoprim–sulfamethoxazole is usually effective, but other antibiotics may sometimes be required.

Importance. Although it is extremely rare, it is important for two reasons. First, it is a treatable cause of chronic CNS inflammation. Second, it is a classical example of how an infection by an unculturable, but widely present pathogen, can cause chronic inflammation in the CNS which is difficult to diagnose.

1.3 Post-Varicella Zoster Virus Cerebellar Syndrome

The sudden onset of problems walking a few days to a few weeks after the onset of chickenpox, i.e., acute Varicella zoster virus (VZV) infection in children, is the usual presentation of post-VZV cerebellar ataxia. Patients develop varying degrees of cerebellar disease usually manifested by ataxia and tremor, sometimes with headache and vomiting. MRI of the brain is usually normal, the CSF can show a pleocytosis, and recovery usually occurs within days to weeks even without any treatment. However, frequently intravenous aciclovir is used. It is unclear whether this disease is infectious or postinfectious; supporting the former is the frequent occurrence of positive VZV DNA in the CSF by PCR. Postviral acute cerebellar ataxia has not been reported with VZV's cousin virus, herpes simplex virus.

Importance. Although this syndrome is uncommon among those with chickenpox, i.e., about 1 in 4,000 cases of chickenpox, chickenpox is such a common infection in the pediatric population that the incidence of this illness is not that low. However, the syndrome is usually short in duration with usually relatively rapid recovery even without therapy.

1.4 Sydenham's Chorea

Sydenham's chorea (SC), first described by Thomas Sydenham in the seventeenth century (see Fig. 16.3) is the most commonly acquired

Fig. 16.3 Thomas Sydenham (1624–1689) was credited with the first descriptions of scarlet fever, caused by Streptococcal infection, and post-streptococcal chorea, also called Sydenham's chorea (figure in public domain at http://commons.wikimedia.org/wiki/Thomas_Sydenham)

chorea of children and presents as involuntary choreiform movements sometimes with neurobehavioral symptoms. Chorea, derived from the Greek word meaning "dance," is an involuntary movement consisting of rapid motions of the hands and feet. SC is considered a parainfectious syndrome because it usually presents a few months after infection with group A β-hemolytic streptococci. Its natural history is that of subacute development to a peak which might last a month or two, followed by spontaneous resolution. The diagnosis of SC is clinical because there are no supportive laboratory studies, although sometimes imaging can reveal inflammation in the basal ganglia. IVIg and plasmapheresis are felt to speed resolution and support the hypothesis that this is an antibody-mediated disease, although this has not been definitively proven.

Importance. Although rare, SC is important because it is an uncommon adverse effect of a very common infection, streptococcal infection. It has been described for years as the classical postinfectious autoantibody-mediated CNS disease, and recent work has provided some molecular support for that pathogenesis [4].

It is also important because it is related to a group of diseases which some clinicians feel are inadequately studied, PANDAS (Pediatric Autoimmune Neuropsychiatric Disorders Associated with Streptococci) .This entity has been described as the underlying disease of a set of children who have neuropsychiatric syndromes such as obsessive–compulsive disorder or involuntary movements following infection with group A-hemolytic streptococci. The link is felt to be an autoimmune reaction associated with antibodies to neurons triggered by the strep infection. The existence of PANDAS is not well accepted and remains controversial.

2   Autoimmune

2.1   Acquired Neuromyotonia

Clinical features of this non-life-threatening and likely underdiagnosed disease are muscle cramps, stiffness, and weakness. Sometimes, myokymia, muscle twitching in the absence of movement of a joint, is a prominent feature, often in the limbs or trunk. Electromyography (EMG) is dramatically abnormal with substantial spontaneous activity, i.e., muscle electrical activity in the absence of attempted movement of the muscle. The disease is thought to be autoimmune because of its association with other autoimmune disease, and because of a positive association with thymoma, a benign thymic tumor. NMT can occur in association with symptoms referable to involvement of the autonomic nervous system such as constipation, excessive salivation and sweating, and cardiac arrhythmias as well as encephalopathic features such as memory loss or confusion; these more complicated forms of NMT are sometimes called Morvan's syndrome.

Recent work has implicated autoantibodies as important in the pathogenesis of NMT. Antibodies to human voltage-gated potassium channels (VGKCs) detected by an immunoprecipitation assay using labeled dendrotoxin can be detected in about 40% of NMT patients (see Chap. 17 for more information on dendrotoxin). Passive transfer (see Chap. 8) of immunoglobulin G from patients with NMT into mice results in increased packets of ACh released and increased neuronal activity in dorsal root ganglion cultures [5].

Importance. NMT is important because it is one of an increasing number of neuroimmunological diseases likely mediated by autoantibodies. Many patients with this disease can now be detected by a new immunoassay made possible by basic research into animal toxins, e.g., dendrotoxin (see descriptions of VGKCs and dendrotoxin in Chaps. 17 and 18).

2.2   Stiff Person Syndrome

Stiff person syndrome (SPS) is a disease caused by excessive muscle contraction of both agonist and antagonist muscles, thought to be due to inhibition of presynaptic inhibitory molecules in the spinal cord, possibly related to the inhibitory neurotransmitter, γ-aminobutyric acid (GABA). Most patients develop the disease spontaneously, but about 5% of the time it occurs as a paraneoplastic syndrome. It is thought to be autoantibody-mediated and autoantibodies against a variety of molecules have been identified including glutamic acid decarboxylase (GAD), amphiphysin, and gephyrin. Patients with SPS have two groups of symptoms (1) muscle stiffness and (2) episodic painful muscle spasms. The clinical diagnosis can be confirmed by EMG documenting simultaneous contractions of agonist and antagonist muscles. Treatment has been with GABA agonists, sedative drugs, and immunotherapies which have been variably successful.

Importance. This syndrome is increasingly being studied because of its association with antibody reactivity against a variety of antigens, and because of its occasional occurrence as a paraneoplastic syndrome.

2.3   Swine-Worker's Neuropathy

From 2006 to 2010, Lachance and colleagues identified 24 patients with neuropathies who worked in swine abattoirs in Minnesota and Indiana. These patients developed a polyradiculoneuropathy, i.e., involvement of many nerve roots and peripheral nerves, sometimes with symptoms referable to the brain or spinal cord in addition. This disease is felt to be due to exposure to aerosolized porcine brain. All of the affected workers had antibodies to mouse brain by immunohistochemistry, and most had IgG antibodies to myelin basic protein (MBP) and an elevated CSF protein. Approximately one-third of workers at the same abattoirs who had no symptoms had anti-brain antibodies. The most involved areas, based on EMG/nerve conduction velocity testing, were the most proximal areas of the nerve, the nerve roots, and the most distal. The patients responded to cessation of exposure and to immunosuppressive medication.

Importance. Although rare, this recently described syndrome demonstrates a number of important points. First, autoimmunity can be induced in susceptible normal humans by repeated exposure to molecules immunologically resembling autoantigens. Second, autoantibodies were detectable in fully one-third of the 85 exposed asymptomatic individuals but not in normal controls. Third, only a very small percentage of exposed individuals became clinically ill. These observations are likely true for many diseases related to the immune system: despite the exposure of many individuals to a particular process, only a fraction of those exposed develop biomarkers consistent with some perturbation, and only a fraction of those with the biomarkers get clinical disease.

3   Idiopathic

3.1   Hashimoto's Encephalopathy

Hashimoto's disease, an autoimmune disease associated with anti-thyroid antibodies, is the most common cause of hypothyroidism in populations who receive adequate dietary iodine and was first described by Hakaru Hashimoto (1881–1934). Dr. Hashimoto left Japan in 1907 to do research in Germany. He focused on the thyroid and found evidence of lymphocytic infiltration in the thyroid of some patients with thyroid disease, what is now called Hashimoto's thyroiditis. It is the most common cause of hypothyroidism, affecting up to 0.5% of the population mostly women. His publication in 1912 on the thyroid, in the *Archiv Fur Klinishe Chirurgie*, the German journal of clinical surgery, was his first and only publication dealing with the thyroid. The term "Hashimoto's encephalopathy" was first used by Lord Brain, a prominent British neurologist, who, in 1966, described a patient with neurological involvement with fluctuating thyroid hormone levels. The term has continued to be used, but is sometimes confused with hypothyroid encephalopathy, in which patients with hypothyroidism from any cause develop neurological symptoms, usually neuropsychiatric manifestations such as cognitive dysfunction, affective disorders, and psychosis, sometimes referred to as "myxedema madness." In contrast, the hallmark of Hashimoto's encephalopathy, as it is usually defined, is its responsive to corticosteroids, implying that the inflammation present in the thyroid of these patients is also present in their brains. Thus, the syndrome has also been called steroid-responsive encephalopathy associated with autoimmune thyroiditis (SREAT) or nonvasculitic autoimmune inflammatory meningoencephalitis (NAIM). It usually presents with encephalopathic symptoms such as behavioral changes, seizures, tremor, or cognitive problems in a previously healthy individual in the absence of structural disease on imaging of the brain or CSF abnormalities. Hashimoto's encephalopathy has been recently reviewed [6]. One of the authors' conclusions is that "despite the many advances we have made in the past 130 years, the fundamental questions in thyroid-related encephalopathies clearly have not changed."

Importance. Some inflammatory syndromes are unusual and overlap with hormonal diseases, making them difficult to precisely define.

Hashimoto's encephalopathy is one of those syndromes and makes corticosteroid therapy an important potential treatment option in patients with undiagnosed encephalopathies.

3.2 Behcet's Syndrome

This syndrome, rare in the USA but more common in Turkey and the eastern Mediterranean area, is primarily a vasculitis. Patients develop recurrent oral ulcerations, and clinical diagnosis requires at least one of the following other manifestations: skin lesions, eye lesions, or a positive pathergy test. Pathergy is the development of a papule 1 or 2 days after a needle prick. There are no diagnostic laboratory biomarkers. Because Behcet's does not have female predominance and because no autoantibody has been defined, some do not feel it is autoimmune in pathogenesis. However, biopsies of affected tissues consistently show considerable mononuclear cell infiltration. Neurological involvement is usually of two different forms, not occurring together in the same patient: the more dangerous first and more common form is parenchymal involvement usually in the brainstem leading to cranial nerve palsies and hemiparesis, while cerebral venous sinus thrombosis leading to headache and raised intracranial pressure is the second, less common, form [7].

Importance. Prevalence rates for Behcet's syndrome in Turkey have been quoted as being as high as 0.4% of the population. Thus, in that part of the world, neuro-Behcet's must frequently be in the differential of CNS inflammatory events. Behcet's resembles Hashimoto's in representing neurological inflammation in a disease that is usually restricted to other tissues: in Behcet's the mucous membranes and eyes, and in Hashimoto's, the thyroid.

3.3 Rasmussen's Encephalitis

Also known as chronic focal encephalitis (CFE), this disease usually presents as a severe acute focal seizure disorder in children. It is important because of its distinctive, dramatic presentation and highly focal severe inflammation. Imaging reveals swelling in one hemisphere with no involvement of the other. The pathology reveals marked infiltration of the brain with mononuclear cells, predominantly with T-lymphocytes, without a clear infectious or autoimmune etiology. Some have hypothesized that the condition is caused by a viral infection but no virus has been isolated. Patients are treated with immunosuppressives but frequently lack of therapeutic response requires hemispherectomy, the surgical excision of one complete hemisphere of the brain.

3.4 Acute Multifocal Placoid Pigment Epitheliopathy

This is a primarily ophthalmological disease causing decreased visual acuity frequently following an acute infectious disease and is diagnosed based on characteristic creamy-white, multifocal lesions in the retina. Neurological involvement in AMPPE is not infrequent and was recently reviewed [8]. Headache or stroke is common, as well as periventricular white matter lesions on MRI of the brain, and the CSF can be consistent with inflammation. The underlying pathogenesis appears to be vasculitis and severe CNS vasculitis can be fatal.

Importance. This disorder underscores the important link between the eye and the brain. Most neuroimmunological diseases can affect the visual system, but some, such as AMPPE, and its cousin, Susac's syndrome (Chap. 6), are primarily ophthalmological diseases which can affect the nervous system.

References

1. Garg RK. Subacute sclerosing panencephalitis. J Neurol. 2008;255:1861–71.
2. Burgoon MP, Caldas YA, Keays KM, Yu X, Gilden DH, Owens GP. Recombinant antibodies generated from both clonal and less abundant plasma cell immunoglobulin G sequences in subacute sclerosing panencephalitis brain are directed against measles virus. J Neurovirol. 2006;12(5):398–402.

3. Panegyres PK, Edis R, Beaman M, Fallon M. Primary Whipple's disease of the brain: characterization of the clinical syndrome and molecular diagnosis. QJM. 2006;99(9):609–23.

4. Kirvan CA, Swedo SE, Heuser JS, Cunningham MW. Mimicry and autoantibody-mediated neuronal cell signaling in Sydenham chorea. Nat Med. 2003;9(7): 914–20.

5. Vincent A. Immunology of disorders of neuromuscular transmission. Acta Neurol Scand Suppl. 2006;183:1–7.

6. Schiess N, Pardo CA. Hashimoto's encephalopathy. Ann N Y Acad Sci. 2008;1142:254–65.

7. Siva A, Saip S. The spectrum of nervous system involvement in Behcet's syndrome and its differential diagnosis. J Neurol. 2009;256(4):513–29.

8. O'Halloran HS, Berger JR, Lee WB, et al. Acute multifocal placoid pigment epitheliopathy and central nervous system involvement: nine new cases and a review of the literature. Ophthalmology. 2001;108(5): 861–8.

Neuroimmunological Molecules and Translational Medicine

17

1 Moving from Basic Science Discoveries to FDA-Approved Therapies

Basic neuroscience is a rapidly moving field, and the nervous system is a vast frontier only a small fraction of which has been explored. The potential for discovery of new molecules and concepts in basic neuroscience to be directly applied to the treatment of human neurological disease is high. However, this potential is not fulfilled as often as possible for a multitude of reasons, some of which were dramatized in the recent popular movie "Extraordinary Measures" in which a new enzyme replacement therapy for Pompe's disease, a glycogen storage disease with dramatic CNS complications, went through a number of hurdles from bench to bedside.

The enzyme replacement therapy depicted in the movie "Extraordinary Measures" is relatively straightforward compared to the highly complex molecules currently being tested for their effects on the immune and nervous systems in neuroimmunological diseases. These agents must be proven to be both safe and effective. Increasingly, new drugs are also being required to be better than currently available therapies, which means that randomized clinical trials (RCTs—see Chap. 7) need to show that the new agent is significantly better than a current therapy, not just better than an inactive placebo.

Another major hurdle for new drugs is that the drug must be demonstrated to be profitable for the pharmaceutical company to develop and sell. Most neuroimmunological diseases are relatively small markets, even multiple sclerosis, the most common neuroimmunological disease. That is why most of the drugs used in neuroimmunological diseases have been developed for other more common diseases, such as rheumatoid arthritis or transplant rejection, and subsequently tested in neuroimmunological diseases after already demonstrated to be successful in a more common disease. Ideally, pharmaceutical companies, nonprofit organizations, and the government would collaborate to bring new agents to the bedside; this does happen to some extent, but in relatively disorganized, random, inefficient, and opaque ways. Over the last 5 years, there has been a substantial decline in new drug approvals, which, coupled with impending loss of patent protection for many profitable medications, have resulted in significant cutbacks in basic research. These factors may exacerbate the inherent difficulties in bringing safe, effective medications to the clinic. In contrast, the advent of the genetics explosion, biologics, and high-throughput screening methodologies have increased the potential to develop and test new agents.

2 Biological Molecules as Therapies

2.1 Therapeutic Monoclonal Antibodies

Therapy of human diseases, including neuroimmunological diseases, has been transformed by the advent of therapeutic monoclonal antibodies

A.R. Pachner, *A Primer of Neuroimmunological Disease*,
DOI 10.1007/978-1-4614-2188-7_17, © Springer Science+Business Media, LLC 2012

(mcAbs) over the past 20 years. The advent of mcAbs was in the mid-1970s when Georges Kohler and Cesar Milstein worked together to develop a technique for producing long-term malignant cell lines producing mcAbs with a defined specificity (see Inset 17.1). The success of this methodology was based on Niels Jerne's natural selection theory of antibody formation which postulated that foreign antigens select the antibody molecules already present within the body which have the best fit.

Inset 17.1 The Scientists Behind Monoclonal Antibodies

Cesar Milstein, born, raised, and educated in Argentina, was a scientist in his late 40s, working at Cambridge University in England, when in 1975 he published his work on the generation of mcAbs using fused hybridoma cells of predetermined specificities; the coauthor in that paper was Georges Kohler, a postdoctoral student from Germany in his late 20s. This methodology was based on the "natural selection" theory of antibody formation formulated by Neils K. Jerne, a Danish immunologist, when he was a research fellow at the California Institute of Technology. For their work, which paved the way for the mcAbs as therapies, and for our current understanding of antibody generation in vivo, Milstein, Kohler, and Jerne shared the Nobel Prize for Medicine in 1984.

The first mcAb approved by the FDA was rituximab in 1997, for B-cell non-Hodgkin lymphoma resistant to other chemotherapies. It is now being used for other lymphomas in addition. Rituximab targets CD20, a molecule on the surface of B cells. Interestingly, the precise mechanisms by which Rituximab is effective in lymphoma therapy are not known, although many possible mechanisms are postulated. Rituximab and other therapies targeting B cells are discussed at greater length in Sect. 2.2.1 as an important neuroimmunological molecule.

- Some of the mcAbs currently being tested for use in MS were FDA-approved for cancer or transplant rejection and are now being tested in MS. Some of these, other than rituximab, are alemtuzumab and daclizumab, which are FDA-approved for B-cell leukemia and for kidney transplant rejection. Natalizumab, which targets α-4 integrin, was developed for both inflammatory bowel disease and MS, and is discussed in greater length in Chap. 7. As with Rituximab, the mechanism of action of these mcAbs is not well understood, but appears to be through a combination of Fc-mediated effector functions once the Fab has bound to the appropriate target. These include antibody-mediated cellular cytotoxicity, complement-dependent cytotoxicity, induction of apoptosis of the target cell, and other effects.

- The efficacy and side effects of mcAbs are primarily defined by the location, distribution, and function of their target. Thus, rituximab and other CD20 mcAbs do not directly affect T cells, macrophages, or monocytes because CD20 is only expressed on B cells. The specificity of mcAbs is, therefore, exquisite and fulfills what Paul Ehrlich (1854–1954), an early twentieth century German immunologist, dreamed of as a "magic bullet." Ehrlich, who won the Nobel prize for the discovery of Salvarsan therapy of syphilis, dreamed of discovering "magic bullets" to treat infectious diseases. These agents would be molecules that would bind to infectious agents with high specificity and toxins could be attached on another end of the molecule to kill the pathogen. Current therapy with mcAbs fulfills some of Ehrlich's hopes about "magic bullets."

- McAbs potentially avoid idiosyncratic organ injuries of chemicals which may only be found either in phase 3 clinical trials or after widespread use. However, targeting a specific molecule with mcAbs may not be completely safe, because of the important role some of these molecules may play in normal functions. For example, a small percentage of patients who receive natalizumab therapy develop progressive multifocal leukoencephalopathy (PML) (see Chap. 7), an unanticipated and frequently deadly consequence of targeting an adhesion molecule.

- Given the "magic bullet" appeal of mcAbs, and their efficacy in treating a number of diseases, it is likely that more and more mcAbs will be developed in the next decades for the treatment of neuroimmunological diseases. Some of these may be developed specifically for MS, the most prevalent neuroimmunological disease, but it is likely that most will continue to be tested in neuroimmunological diseases only after having been shown to be effective in cancer, autoimmune problems, or other diseases. It is also likely that unforeseen side effects will continue to occur, since our knowledge of the function and distribution of potential targets will frequently lag behind the eagerness to use these "magic bullets" in our war against disease.

2.1.1 Biologics

mcAbs represent a member of a class of relatively new therapies called biologics. These are materials isolated from natural sources and are usually produced by biotechnologies such as molecular or cellular biological techniques, usually using recombinant DNA methodology. Thus, they are differentiated from chemicals, which historically have been the major source of therapeutic drugs. Usually, biologics are regulated differently from chemicals, because of major differences in how they are produced and in their quality control.

The biologics currently most frequently used in neuroimmunological diseases are the interferon-β compounds in MS. These are produced either by bacteria (IFN-β-1b) or by Chinese hamster ovary cells (IFN-β-1a) and are discussed in more detail in Chap. 7. Another form of biologic widely used in neuroimmunological disease is intravenous immunoglobulin (IVIg), which is discussed in Chap. 19 in greater length.

Another example of a type of biologic treatment that can be used to specifically target a particular molecule is a recombinant decoy receptor, an example of which is etanercept. This drug, approved for use by the FDA for psoriasis and various forms of arthritis, utilizes the known structure of one of the tumor necrosis factor (TNF) receptors to bind to soluble TNF-α, a pro-inflammatory cytokine. Etanercept blocks TNF-α action by binding to TNF in the serum and preventing its binding to TNF receptors in the body. It is a fusion molecule consisting of the Fc portion of human IgG1 and the p75 portion of the TNF-α receptor. No decoy receptor biologic is yet available for patients with neuroimmunological disease.

2.2 Molecules as Targets of Therapy or of Disease: CD20, Aquaporin-4 Receptor, TNF-α, Sphingosine-1-Phosphate Receptors, CD52, MOG, Channels

2.2.1 CD20

(a) *The molecule.* CD20 is a protein of unknown function present on the surface of nearly all B cells, with the exception of very early B cells and fully differentiated plasma cells. It does not appear to be present on other cell types.

(b) *The target.* The first mcAb to be FDA-approved, rituximab, targeted CD20 for the treatment of lymphoma; this therapy has undergone some testing in patients with MS, but further studies on anti-CD20 therapies in MS will likely use molecules other than rituximab. One reason for the need for other forms of anti-CD20 was that some MS patients developed anti-rituximab antibodies. Thus, other more "humanized" anti-CD20 mcAbs, such as ocrelizumab and ofatumumab, are currently being tested in MS clinical trials.

2.2.2 Aquaporin-4

1. *The molecule.* The aquaporins are a group of membrane molecules whose functions include regulation of the flow of water through cells. Aquaporin is a complex molecule (Fig. 17.1) with a number of transmembrane loops and intracellular and extracellular domains. There are at least four different aquaporin molecules. Aquaporins were first described by Peter Agre (see Inset 17.2). Aquaporin-4 in the CNS is localized to astrocytic end-feet (Fig. 18.1).

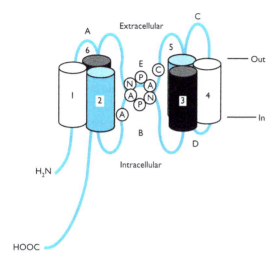

Fig. 17.1 Hypothetical structure of the aquaporin water channels. The water channels consist of six transmembrane-spanning regions (*cylinders*), which surround two loops of the protein (E and B) that form a pore of just the right size in the membrane to allow passage of water molecules. Because this is a passive transport process, water can flow in either direction across the membrane, depending on concentration gradients (out=extracellular, in=intracellular). Other neutrally charged molecules of a similar diameter (e.g., urea) also can pass through this channel. This hypothetical structure recently has been confirmed by X-ray crystallography studies

---

Inset 17.2 Peter Agre and Aquaporin

Dr. Agre's discovery of aquaporin was serendipitous. In 1988 while attempting to purify the 30 kDa Rh, a protein on the surface of red blood cells, his laboratory was faced with a troublesome 28 kDa "contaminant," initially thought to be a fragment of the Rh protein. His work ultimately shifted from work on the Rh protein to work on the 28 kDa protein which proved to be a water channel protein, which he called "aquaporin."

---

2. *The target*. One of the most notable advances in the last decade in neuroimmunological disease has been the finding that the serum of a large percentage of patients with Devic's disease, also called neuromyelitis optica (NMO), contains autoantibodies which target aquaporin-4 in the CNS. These autoantibodies

appear to be pathogenic in animal models of the disease, but the link between these autoantibodies and the clinical phenotype of the disease remains obscure. NMO is a mimic of multiple sclerosis (see Chap. 6), and anti-aquaporin-4 antibodies are generally not present in MS. Since NMO is treated differently than is MS, the identification of aquaporin-4 as a target of autoantibodies has spurred the development and use of a test for these antibodies (see Chap. 18) in patients with consistent clinical pictures of CNS inflammation and demyelination, which has aided in the use of appropriate treatment for NMO.

2.2.3   Tumor Necrosis-α

1. *The molecule*. TNF-α is about 17 kDa in size and usually is studied as a soluble cytokine. It is produced predominantly by macrophages during inflammatory reactions and binds to well-characterized TNF receptors which are present on most cells. A local increase in TNF will result in the clinical hallmarks of inflammation. It is structurally related to TNF-β, another cytokine, also called lymphotoxin.

2. *The target*. The highly inflammatory nature of TNF-α has made it into one of the most popular targets for anti-inflammatory therapies. The mcAbs infliximab, certolizumab, and adalimumab, as well as the fusion protein, etanercept, specifically target TNF-α. These biologics are used in a variety of inflammatory diseases, but the most prevalent use is in rheumatoid arthritis, a relatively common idiopathic condition, affecting 1% of the population, causing pain, swelling, and deformity in the joints.

   Early in the development of anti-TNF-α biologics, MS seemed an obvious disease to which to apply these therapies. Investigators found that TNF-α blockade in the animal model of MS, experimental autoimmune encephalomyelitis (EAE) (see Chap. 8) resulted in marked improvement in disease, providing further apparent rationale for its use. CSF levels of TNF-α correlated with disease progression. However, a number of studies, including a relatively large phase II study, demonstrated that not only was TNF blockade not helpful, but also it could potentially worsen MS relative to untreated controls [1]. The mechanisms by which TNF blockade ameliorates inflammation in rheumatoid arthritis, but exacerbates it in MS, are unclear. There are two important take-home messages: first, that inflammation in the CNS is not necessarily responsive to the same manipulations as inflammation outside of the CNS, and second, that results in the EAE model of MS are frequently not predictive of results in the human disease.

2.2.4 Sphingosine-1 Phosphate Receptor

1. *The molecule.* Sphingosine-1 phosphate (S1P) receptors represent a family of receptors that bind S1P and are part of a broader family called high-affinity G protein-coupled receptors (GPCR). S1P is made from intracellular sphingosine during normal cell maintenance of the sphingomyelin present in the cell membrane. Levels of this molecule are high in plasma but low intracellularly; S1P levels also increase intracellularly in sites of inflammation.

2. *The target.* Because of the above facts, S1P receptors were felt to be potential targets for immunosuppression. The drug fingolimod (see Chap. 7), structurally related to S1P, was found to be an S1P functional antagonist, working by down-modulation of lymphocytic S1P to slow S1P-dependent egress of lymphocytes, especially T cells, out of lymph nodes, and thus resulting in substantial lymphopenia in the peripheral blood. It was tested for its effect on organ transplantation, but was not more effective than currently available therapy. However, recent studies in MS showed substantial efficacy with reasonable safety, and the drug has been FDA-approved for use in MS (see Chap. 19).

2.2.5 CD52

1. *The molecule.* CD52 is a protein present in large amounts on the surface of mature lymphocytes, monocytes, and dendritic cells but not on their stem cells. Thus, CD52 treatment with the drug alemtuzumab depletes these cells but the depletion is not permanent.

Interestingly, CD52 is not present on mouse lymphocytes, so functional experiments are hard to design, and the functional role of CD52 is unknown.

2. *The target.* Alemtuzumab is FDA-approved for use as second-line therapy for chronic lymphocytic leukemia. Because CD52 is present on such a large percentage of cells, alemtuzumab results in marked immunosuppression and significant adverse effects. In the recent phase II study of alemtuzumab in MS, there was one fatal case of immune thrombocytopenic purpura (ITP), a high incidence of autoimmune thyroiditis, and a higher incidence of infections [2].

2.2.6 MOG

1. *The molecule.* Myelin oligodendrocyte glycoprotein (MOG) is an approximately 25 kDa sized protein which has attracted a great deal of attention as a potential autoantigen in MS. It has been utilized extensively as an encephalitogen, i.e., inducer of CNS injury when injected into experimental animals in EAE as described in Chap. 7. It is exciting for at least three reasons. First, it is found exclusively in CNS myelin. Second, it is one of the only proteins of myelin that is expressed extracellularly, on the outermost lamella of the myelin sheet, and thus would be a logical target of autoantibodies and cellular immunity. Third, the full-length protein as well as certain peptides of the protein are highly active in multiple different animal models of EAE.

2. *The target.* In most models of EAE induced by MOG, anti-MOG antibody production is necessary to induce full-blown disease including inflammation, demyelination, and substantial disability. Some data supports the hypothesis that early in MOG-induced EAE, T cells with specificity for MOG result in inflammation within the CNS making the blood–brain barrier more porous and that later anti-MOG antibodies diffuse through the impaired blood–brain barrier and result in antibody-mediated injury demyelination and injury. This two step hypothesis is supported by the finding that monoclonal anti-MOG antibodies

Table 17.1 Autoimmune channelopathies

Targeted channel	Clinical phenotype	Clinical assay
Nicotinic acetylcholine receptor (nAChR)	Myasthenia gravis	Utilizes iodinated bungarotoxin and immunoprecipitation
Voltage-gated calcium channel (VGCC)	Lambert–Eaton myasthenic syndrome	Utilizes iodinated conotoxin and immunoprecipitation
Aquaporin receptor	Neuromyelitis optica	Immunofluorescence of brain slices; receptor transfected cell
Voltage-gated potassium channel (VGKC)	Limbic encephalitis; neuromyotonia	Utilizes iodinated dendrotoxin and immunoprecipitation
N-methyl-D-aspartate (NMDA) receptor	Limbic encephalitis	Immunofluorescence of brain slices; receptor transfected cell

are able to cause demyelination both in vivo and in vitro, and can induce significant disease in B-cell deficient mice in whom the blood–brain barrier is breached. The role of anti-MOG antibodies is less clear in human multiple sclerosis; there have been reports of both the importance of these antibodies [3] and their lack of importance [4].

2.2.7 Channels as Targets of Autoantibodies

An exploding area of research, discussed briefly in Chaps. 10 and 14, is the elucidation of autoantibodies to channels in neuroimmunological disease. The prototype autoimmune channelopathy is, of course, myasthenia gravis, caused by autoantibodies to the acetylcholine receptor (AChR) and discussed in Chap. 10. However, a range of others have been described over the last 10 years (see Table 17.1), and their mechanisms of action are actively being investigated. Most of them are felt to result in dysfunction by receptor modulation, cross-linking or internalization, thus mostly independent of immune effector mechanisms such as complement-mediated destruction or antibody-dependent cellular cytotoxicity. The autoantibodies in these diseases are directed against parts of these receptors that are outside the cell, providing further evidence that the antibodies are indeed pathogenic.

Understanding of these channels and antibodies to them has been advanced greatly by improved availability of a variety of animal toxins summarized briefly in Chap. 10. Dendrotoxin comes from the venom of the mamba snake

(Dendroaspis), one of the deadliest snakes in the world, and works by blocking voltage-gated potassium channels (VGKCs). Since dendrotoxin binds with high affinity to VGKCs, it can be used to label VGKCs in immunoassays for anti-VGKC antibodies, similar to the use of the krait toxin, bungarotoxin, in labeling AChRs for anti-AChR antibody testing for patients with myasthenia gravis (see Chap. 10). The use of animal toxins for assaying for anti-channel and -receptor antibodies is described in greater depth in Chap. 18. These toxins have been utilized primarily for attempts to understand the structure and function of channels and receptors. Their utilization for identifying autoantibodies has been an unanticipated, but welcome, offshoot of this research into basic neuroscience. It represents an excellent example of how basic research, seemingly unrelated to any clinical issues, can have direct relevance to disease.

2.3 Neuroprotection: The Holy Grail of Translational Neurology

Many neuroimmunological diseases, particularly MS, result in marked and progressive neuronal and myelin loss. These diseases would benefit from the development of neuroprotective drugs, defined as agents which could ameliorate the downhill progression of cell loss. Thus, a few years ago, neuroimmunologists were excited by a large neuroprotection study of NXY-059 in acute ischemic stroke. This molecule, called a "spin trap" molecule, was effective in in vitro systems in scavenging free radicals, which were felt to be important in accelerating ischemia and injury. Substantial in vitro and animal model work had been done to justify large, very expensive human studies on this molecule. Guidelines for preclinical studies had been developed in 1999 by the Stroke Therapy Academic Industry Roundtable (STAIR) which was anticipated to assist the pharmaceutical industry in bringing drugs with a high probability of success to phase 3 studies. In addition, stroke seemed to be an ideal choice for neuroprotection studies since in stroke there is sudden severe damage, and the prediction was that since the drug should have a substantial effect, this treatment effect should have been readily seen in a short period of time. Unfortunately, in two very large studies, recruiting 5,208 patients, the drug was proven to be ineffective in humans [5], prompting much hand-wringing and attempts at explanations of why preclinical studies had not predicted the situation in humans. The STAIR recommendations were felt to have not been "closely followed."

This trial and the attempts to develop neuroprotective drugs in Alzheimer's disease, stroke, Parkinson's disease, and other neurological "degenerative" diseases have direct relevance to neuroimmunology. There have been extensive efforts to identify a neuroprotective agent in the most common and debilitating neuroimmunological disease, multiple sclerosis, but no attractive target has yet risen clearly above numerous potential molecules and processes. Current therapy of MS appears to have little or no effect on progressive damage to the CNS (for more information on this topic, see Chap. 7). The need to develop safe and effective neuroprotective agents for MS is likely the greatest challenge to research neuroimmunologists in the near future. The following statement of STAIR is sobering and applies equally well to MS as it does for stroke:

> The costs of modern drug discovery and development in most therapeutic disciplines have become almost prohibitively expensive for any pharmaceutical company, because investments in a single drug approximate $0.5 to $1.0 billion in Research and Development activities. The bounty of new molecular targets derived from the recently unraveled human genome poses unprecedented challenges to define and validate the biology, pathology, and clinical usefulness of these targets for safe and effective drug development. These new realities, increased by regulatory requirements for novelty and differentiation of new therapies over those currently available, creates unprecedented hurdles in developing new drugs. [6]

References

1. TNF neutralization in MS: results of a randomized, placebo-controlled multicenter study. The Lenercept Multiple Sclerosis Study Group and The University of British Columbia MS/MRI Analysis Group. Neurology. 1999;53(3):457–65.
2. Coles AJ, Compston DA, Selmaj KW, et al. Alemtuzumab vs. interferon beta-1a in early multiple sclerosis. N Engl J Med. 2008;359(17):1786–801.
3. O'Connor KC, Appel H, Bregoli L, et al. Antibodies from inflamed central nervous system tissue recognize myelin oligodendrocyte glycoprotein. J Immunol. 2005;175(3):1974–82.
4. Owens GP, Bennett JL, Lassmann H, et al. Antibodies produced by clonally expanded plasma cells in multiple sclerosis cerebrospinal fluid. Ann Neurol. 2009;65(6):639–49.
5. Diener HC, Lees KR, Lyden P, et al. NXY-059 for the treatment of acute stroke: pooled analysis of the SAINT I and II Trials. Stroke. 2008;39(6):1751–8.
6. Fisher M, Feuerstein G, Howells DW, et al. Update of the stroke therapy academic industry roundtable preclinical recommendations. Stroke. 2009;40(6):2244–50.

Neuroimmunological Diagnostic Tests

18

This chapter discusses diagnostic tests commonly used by neuroimmunologists which are specialized for use in inflammatory and infectious diseases. These are generally assays of the blood or cerebrospinal fluid (CSF), two bodily fluids readily accessible to the clinician. We will not discuss the standard neurological tests performed by neurologists in evaluating neurological disease, such as the history, examination, imaging of the brain by MRI or CT scanning, electroencephalography, electromyography, nerve conduction studies, or evoked responses, which are discussed in Chap. 2. Consistent with its strong link to basic research, much of clinical neuroimmunology is dependent on laboratory assays. Researchers are constantly trying to move assays developed in research laboratories into the clinical arena, albeit with varying degrees of success. As we have discussed concerning the value of surrogate markers of diseases, it is one thing to be able to demonstrate statistically robust sensitivity and specificity of a test for some aspect of disease, but it is another thing entirely to establish the clinical relevance of that finding. This means the clinical neuroimmunologist must have a firm understanding of the laboratory assays used in the field.

1 General Considerations for Testing

Tests are obtained after history and physical examinations are performed to gain further information after a differential diagnosis has been generated from the history and examination. Usually, tests are done to confirm diagnoses high on the differential diagnosis. There has been an increase in diagnostic testing over the past 30 years, most of it appropriate, but much of it inappropriate and excessive. A test should not be ordered as an automatic, knee-jerk reaction. Tests should be utilized according to their sensitivity, specificity, and predictive value within a given population. Sensitivity is defined as how likely the test will be positive in an affected individual; specificity is how likely the test will be negative in an unaffected individual; and predictive value is how the test will perform in a certain population. This means if a test has a high sensitivity, high specificity, and it is used in a population which has a reasonably high likelihood of being affected, the predictive value of either a positive test or negative test will be very high. An example is PCR of the CSF for herpes simplex genome. Because this test has high sensitivity and specificity, a positive HSV PCR has high predictive value in a population of patients with fever, seizures, and impaired level of consciousness; i.e., essentially, 100% of test-positive patients in this population will have active herpes simplex viral infection as a cause of their symptoms. In contrast, a *Borrelia burgdorferi* serum ELISA (in the absence of immunoblotting, see below), which has high sensitivity but low specificity in a population of patients in an endemic area with nonspecific symptoms, has poor predictive value. That is, a very low percentage of individuals in a population with nonspecific symptoms and positive

Borrelia burgdorferi serum antibody tests will have active infection with *Borrelia burgdorferi* as a cause of their symptoms. However, in a cohort which is better defined for clinical characteristics of Lyme disease, the specificity may be quite high. The usefulness of a test is dependent not only on its capacity to measure the surrogate marker but also on the way that the relevant population to be tested with it is defined. Tests should always be interpreted in the context of their sensitivity, specificity, and predictive value. Often this puts the onus on the neuroimmunologist to be able to define a given test's usefulness in varying clinical situations.

1.1   Validation of Laboratory Tests by Regulatory Authorities

In the USA, authorities have been mandated to review and approve the performance of laboratory tests that might aid in diagnosis or therapy of a particular disease. While this does not result in a ruling on the clinical usefulness of a given test, it does assure that the process of running the assay are subject to special quality control standards before they are allowed to be utilized in clinical care. In the USA, this process is regulated by the federal and state governments under the CLIA program (Clinical Laboratory Improvement Amendments) aided by the College of American Pathologists. An example is the anti-JC virus antibody assay which will likely aid in the management of patients with MS being considered for natalizumab therapy. This test, developed and currently being run in a research setting, has just recently been approved for clinical use, and has become routinely available to clinicians (see discussion of JCV antibody in this chapter). After new tests become CLIA certified, and available to clinicians, another level of analysis occurs, as clinicians determine the assay performance in a broad population of patients. Within a few years after clinical availability of an assay, groups of clinicians perform studies evaluating the assay's sensitivity, specificity, and predictive value in patient populations and publish the studies in the medical literature. Over time the new test takes a place in clinical practice.

2   Blood Tests

Blood is a body fluid readily accessed by venipuncture, the procedure by which a needle is inserted into the vein and blood is withdrawn. The blood is evacuated into tubes which may contain various materials which initiate the processing of the sample in various ways. In the absence of the milieu of the vascular compartment, once the blood enters the glass tube, platelets and thrombin cascade elements in the blood will result in clotting, a thickening and stickiness of blood components. Thus, if the blood is drawn into a tube without any chemicals that would inhibit the coagulation systems of the blood, clotting will begin to occur almost instantly and over minutes to hours the blood will completely clot. The non-clotted portion is called serum; most tests of the noncellular components of the blood test serum. Tests of the blood for the cellular elements, either white or red blood cells, require that the tube into which the blood is drawn contain an anticoagulant such as heparin sulfate, so that the blood does not clot. If anticoagulated blood is separated into cellular versus noncellular fractions, usually by centrifugation, the noncellular fraction is referred to as plasma.

2.1   Blood Tests for Inflammation

Some blood tests are ordered by neuroimmunologists to assess markers in the blood indicative of an inflammatory process which might affect the nervous system. None of these tests are highly sensitive nor specific for inflammation deriving from a particular immune response or a response that defines a particular immunologically mediated disease, and usually serve as general indicators. They are especially helpful in systemic diseases such as the vasculitides or infections that might affect the nervous system. Though these may be targeting a particular organ system or tissue, such diseases often are associated with the release of inflammatory mediators and cells into the vascular compartment, and are thus amenable to assays to detect their presence.

2.1.1 Sedimentation Rate

Sometimes called erythrocyte sedimentation rate (ESR) or Westergren sedimentation rate (WSR), this test, when elevated, indicates that inflammation is present somewhere in the body, often as a generalized inflammatory disease, infection, or malignancy. The sedimentation rate measures how far red blood cells in blood migrate along a glass surface. The precise molecular pathways by which the sedimentation rate are increased in inflammation are unknown. In some inflammatory diseases limited to the CNS, such as multiple sclerosis, the ESR is not generally increased, even when CNS inflammation is extensive. The reasons for this finding are unknown, but one expects that for an ESR blood test to be positive, most likely there must be ready communication between the inflamed area and the blood circulation, and this access is limited by the blood–brain barrier in such diseases of the CNS as MS.

2.1.2 White Blood Cell Count Elevation

A complete blood count (CBC) measures the number and kinds of cells in the blood, and an elevation in the number of white blood cells in the blood usually indicates inflammation.

2.1.3 Surface Markers of White Blood Cells

The neuroimmunologist is frequently interested in determining the possible disease-specific pattern of white blood cell types in the peripheral blood or CSF. The CBC provides a count of neutrophils, lymphocytes, monocytes, eosinophils, and basophils, which are differentiated by size (charge) in automated counters and by morphology on microscopic visual exam. Additional information may be obtained by identifying molecules on the surface of the white cells, called surface markers, and have been categorized nosologically as different cluster of differentiation (CD) molecules. This is especially important for differentiation of T lymphocytes, which express the CD3 antigen, from B lymphocytes, which express CD19 or CD20 and in identifying how many T lymphocytes are "helper"(CD4) relative to "cytotoxic"(CD8), as briefly explained in

Chap. 1. A clinical example of the value of such a test is in HIV: later stages of HIV infection are characterized by a profound depletion in CD4 cells in the blood so that the CD4/CD8 ratio in the blood, which is usually between 1 and 4, is markedly lowered, sometimes to as low as 0.1. Another important clinical use of determining these markers is in the case of lymphoma in the CNS (Chap. 14), since CNS lymphomas are almost without exception, B-cell malignancies, and the demonstration that most of the cells in the CSF have B-cell markers, i.e., either CD19 or CD20, supports the diagnosis of lymphoma.

To determine CD markers in a population of cells, derived either from the blood or CSF, a methodology called fluorescence-activated cell sorting (FACS) is used. After cells are exposed to fluorescence-tagged monoclonal antibodies reactive to a certain CD marker, they are allowed to pass through a narrow tube in single file, and the amount of fluoroscence that each individual cell produces due to the amount of tagged MAB adherent to it is assayed as it passes by a sensor. The population is analyzed for how many cells contain fluorescence, which identifies how many cells have the CD marker. Modern FACS machines can analyze many markers with different fluorescent dyes at the same time, allowing identification of multiple markers on the same cells

2.2 Blood Tests for Specific Immunity to Pathogens or Autoantibodies

Automated technologies and the advent of molecular biology have resulted in markedly improved methodologies for detecting antibodies directed against a wide range of targets. Many antibody assays utilize the ELISA methodology that allows high throughput and rapid results. ELISA methods frequently require confirmation with a second assay such as immunoblotting or western blotting (see Inset 18.1). Another commonly used name for such testing for antibody in the blood when it pertains to immunity to pathogens is serology. Assays for autoantibodies can include any of the above methodologies and also may be immunohistochemical or immunofluorescence assays, such as the aquaporin antibody screen (see below), in which processed tissues are used as the antigen.

Inset 18.1 ELISA/Immunoblotting

Testing for antibodies to *Borrelia burgdorferi* is an example of how testing for antibodies to a pathogen is performed in the clinical laboratory. The word ELISA stands for enzyme-linked immunosorbent assay, in which the assay is done completely in solid phase. There are many modifications of the ELISA; the one most commonly used for detection of antibodies to a pathogen is the indirect ELISA in which, in the case of *B. burgdorferi*-specific antibodies, sonicated spirochetes are incubated in the plastic wells in an alkaline buffer, which leads to adherence of the molecules to the plastic. After washing off unbound molecules, the patient's serum is incubated on the plate, and antibodies with high affinity for *B. burgdoreri* molecules bind to these molecules on the plate. After unbound material is washed off, the antibodies bound to the plate are detected by an anti-immunoglobulin antibody which has been chemically conjugated to an enzyme; this chimeric molecule is often called the ELISA conjugate. After unbound conjugate is washed off, the enzyme moiety of the conjugate remaining bound to the patient's immunoglobulin on the plate reacts with an ELISA substrate, which results in the development of a color. This color reaction can then be read by a specialized spectophotomer, an ELISA reader. Quantitation of anti-*B. burgdorferi* antibody can be performed by quantitation of color developed, since it will be directly proportional to the amount of enzyme, which is directly proportional to the patient's anti-*B. burgdorferi* antibody bound to the plate.

Immunoblotting, sometimes called western blotting, is a technique in which the mixture of proteins of the pathogen is separated on a membrane utilizing a methodology called electrophoresis. In electrophoresis, molecules of complex solutions are separated in solid phases by movement through an electrical field; larger molecules move slower than smaller molecules, and markers can identify the separated molecules, usually proteins, by size. Thus, a "41 kDa band" identifies antibodies binding to a 41,000 molecular weight protein, which in *B. burgdorferi*, is a protein found in the flagella of the spirochete. This flagellin is found in other bacteria, and since humans become exposed to many bacteria, antibody to the 41 kDa protein is a relatively nonspecific antibody population found in a large percentage of the general population. Other proteins, such as the 39 kDa protein are more specific for *B. burgdorferi*. Thus, a positive western blot requires reactivity to multiple *B. burgdorferi* proteins. Most patients with Lyme neuroborreliosis have had prolonged exposure to the spirochete, have developed high levels of antibody, and will react to more than the five bands required for a positive IgG immunoblot.

2.2.1   Serum Antibodies

Pathogens

Detection of antibodies to a pathogen indicates exposure of the immune system to the pathogen at some point in the life of the individual and may not necessarily be able to demonstrate the current, actual presence of the pathogen. Whether exposure and concurrent infection are identical and whether a positive test indicates that the pathogen is causative of the disease process depends on the pathogen.

For instance, positive anti-HIV antibodies almost always indicate the presence of the pathogen, HIV, since an individual who is HIV positive is highly unlikely to ever be able to completely clear the pathogen from the body. In contrast, in Lyme disease, many individuals can continue to be antibody positive for years after extensive antibiotic treatment and complete clearance of the pathogen. In chronic viral infections which are common in the general population such as the JC, herpes, CMV viruses, a positive antibody test indeed indicates active infection, but the active infection is controlled by the body's immune response, and the vast majority of infected individuals have no clinical disease referable to the active infection. For another pathogen, *Clostridium tetani*, the causative agent of tetanus, almost all individuals have detectable serum antibodies to *C. tetani* due to their having received tetanus vaccinations. Of course, none of these individuals has active infection with the bacteria. Thus, in these four infections—HIV, *Borrelia burgdorferi*, JCV, and *Clostridium tetani*—the presence of a positive anti-pathogen antibody assay in the serum means something very different.

Viruses

(i) *HIV*. Improved methodologies have allowed HIV testing to become much easier than they were only a decade ago. OraQuick and similar point-of-care tests are anti-HIV antibody tests that provide results in less than half an hour. Oral fluid is obtained, mixed with a solution, and the results are read from a stick-like testing device. The assay is highly reliable and is used widely as a point of care screening assay. ELISAs and immunoblots are additionally used.

(ii) *JC virus (JCV)*. JCV is the causative agent of progressive multifocal leukoencephalopathy (PML), and increasingly of concern in MS patients treated with a variety of new agents (see Chaps. 7 and 19). The vast majority of individuals who are infected with JCV have asymptomatic medically unimportant infections, well controlled by a normally functioning immune system, but JCV can become reactivated to cause PML (see Chap. 12) if the patient becomes immunosuppressed either by HIV infection or by being treated with immunosuppressive medications.

ELISAs to detect the presence of anti-JCV antibodies have been developed for clinical use, and, as of late 2011, have just become be available to clinicians as routine clinical assays. Thus, a negative JCV antibody assay identifies a patient population at very low risk of developing PML when immunosuppressive drugs such as natalizumab are used, relative to the JCV-positive population. The percentage of adults who are seropositive in the normal population ranges from 60 to 80%, depending on methodology [1].

(iii) *West Nile Virus (WNV)*. WNV is a flavivirus which caused an epidemic of meningoencephalitis in the USA in 2003, and has continued to be associated with sporadic outbreaks. Serology is the diagnostic assay of choice in this infection since the virus is only detectable in the blood for a short time, almost always prior to the onset of symptoms, and by the time the patient is symptomatic the virus has been cleared from the blood. The IgM response is dominant in this infection; persistent IgM positivity can remain for many months after clearance of the encephalitis.

(iv) *Common viruses in the population*. Adults are infected with a variety of viruses, and in many instances, these infections are persistent, leading to persistently positive antibodies in the serum. Examples include Epstein–Barr virus (EBV) and cytomegalovirus (CMV). Although a role has been suggested for EBV in diseases such as chronic fatigue syndrome (CFS) or fibromyalgia, the evidence for this is meager. The evidence for the role of EBV in HIV-associated CNS lymphoma or CMV in HIV-associated polyradiculitis is stronger, but determination of serum antibody lends little to the diagnosis or management of these syndromes. There has been some interest in EBV as a possible pathogen or co-pathogen in multiple sclerosis. At this time, there appears be little role for serum anti-EBV or -CMV testing in neuroimmunological disease.

Bacteria

Most infections with bacteria are diagnosed by culture, but for some so-called fastidious organisms, such as spirochetes (syphilis and *Borrelia burgdorferi*) and mycobacteria, culture is a low-yield assay, and other assays are clinically more important. For spirochetes, serology is the assay of choice, while for *Mycobacteria tuberculosis*, skin testing or PCR are the tests of choice. The difficulties in making the diagnosis of tuberculosis within the CNS, discussed in Chap. 12, are legendary. Even though serology is the preferred means of diagnosis in both neurosyphilis and Lyme neuroborreliosis (LNB), the type of anti-spirochetal antibody tests used for neurosyphilis and neuroborreliosis are very different.

In Lyme neuroborreliosis, an ELISA, coating plates with lysates from sonicates spirochetes grown in culture, is the first step, which, if positive, is followed by an immunoblot (see Inset 18.1, ELISA/immunoblotting).

In contrast, in neurosyphilis, the first fast, inexpensive screening test is a reaginic test, a VDRL or RPR, autoantibody tests described in Chap. 13, based on autoantibodies to phospholipids which develop during infection with *Treponema pallidum*. If this screening assay is negative, there generally is no need to proceed with a second level test. However, if it positive, a specific treponemal antibody assay needs to be performed, since there are many causes of false-positive reaginic antibody tests. These false-positive assays, sometimes referred to a biological false positive (BFP), are frequently seen in systemic lupus erythematosus (SLE) or other autoimmune diseases, as described in Chap. 13, but can also be caused by other infections such as TB, malaria, hepatitis viruses, varicella zoster, or measles, or even by cancers such as lymphoma. The second level tests are antibody assays in which Treponemal material serves as the antigen, e.g., fluorescent treponemal antibody (FTA) or the *Treponema pallidum* hemagglutination assay (TPHA).

Fungi

The diagnosis of CNS fungal infections is ideally performed by culture, or identification of fungal antigen (e.g., cryptococcal antigen, see Chap. 12), but antifungal antibodies in the serum are sometimes helpful, especially in CNS histoplasmosis or coccidoidomycosis. Most fungi, however, can be cultured from the CSF.

Autoantibodies

Autoantibodies, defined as antibodies against host antigens, can be detected in a wide variety of neuroimmunological diseases. Some autoantibodies appear to be pathogenic, while most are biomarkers of an underlying disease process.

Reaginic Antibody Tests for Syphilis and Antiphospholipid Antibodies

As mentioned in Chap. 13, autoantibodies to phospholipids (the VDRL or RPR) reliably appear during syphilis and are excellent markers of infection. Thus, if the infection is treated, the titers of serum VDRL or RPR drop, and usually clear over time. Antiphospholipid antibodies detected by the VDRL or RPR assay or by the anticardiolipin ELISA are also a feature of primary or secondary antiphospholipid antibody syndromes.

Antinuclear Antibodies and Other Autoantibodies in Lupus and Related Rheumatological Disease

In SLE, the autoimmune response generates autoantibodies to the components of the nuclear membranes or nucleoplasm of particular cells, commonly called antinuclear antibodies (ANA). Patients with different types of autoimmune disorders frequently have ANAs but their types are different, since antigens to which they bind are different. For instance, autoantibodies in Sjogren's syndrome, discussed in Chap. 8 as a mimic of MS, are directed against nuclear antigens called Ro and La, sometimes called SS-A and SS-B.

Autoantibodies to Neurological Antigens

This area has attracted increasing interest in the last decade, primarily because of a number of discoveries, including the documentation of autoantibodies to aquaporin-4 receptors in neuromyelitis optica (NMO), and the discovery of a number of new paraneoplastic syndromes that appear to be autoantibody mediated.

Fig. 18.1 Schematic representation of a brain capillary at the blood–brain barrier. Aquaporin-4 is expressed at the end feet processes of the perivascular astrocytes, providing a pathway for brain edema clearance at the level of blood–brain barrier

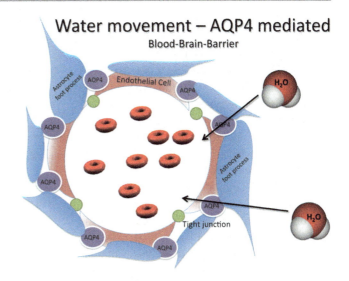

(i) *Aquaporin-4 in NMO.* In NMO, discussed in Chap. 8, antibodies to aquaporin-4 are found in the serum of the majority of patients. Aquaporin-4 (Fig. 17.1), a membrane protein, is localized to the endfeet of astrocytes (Fig. 18.1) and is thought to constitute part of the blood–brain barrier; why this protein has become an antigenic target in NMO is unknown. Patients with NMO have a phenotype of an MS-like disease consisting of extensive and continuous spinal cord lesions, and/or optic neuritis, often have normal brain MRI scans and negative oligoclonal band testing. The serum antibodies are thought to be pathogenic, a hypothesis supported by passive transfer experiments in experimental animals [2], in which antibodies from patients with NMO and positive anti-aquaporin antibodies induced exacerbation of EAE in rats. The antibodies themselves were not pathogenic in the absence of EAE since they did not cross the blood–brain barrier. Severe disease required the presence of both a pathogenic antibody in the serum and breakdown of the blood–brain barrier induced by EAE. The same combination of factors may be operative in human NMO in which both a breakdown in the blood–brain barrier and pathogenic antibodies in the blood may need to be present to result in disease.

Initially, when first described, the assay for aquaporin-4 receptor antibodies was performed by reacting patient's sera with mouse cerebellar slices [3], but now the assays are performed using cell lines transfected with aquaporin-4 receptors. The test is not perfect for defining the clinical syndrome; the sensitivity is 50–80% depending on the laboratory, but the specificity is quite high, in some studies over 90%. Thus, a positive test usually indicates that the patient has NMO, but a negative test does not rule out the diagnosis.

(ii) *AChR.* The measurement of anti-AChR antibodies in the serum is a very helpful diagnostic assay for myasthenia gravis. The methodology used is generally an immunoprecipitation assay, in which anti-AChR antibodies in the serum of MG patients bind to and precipitate in vitro AChR from muscle homogenates after the AChR has been labeled with radioactive bungarotoxin (see Chap. 10 for information about bungarotoxin). This immunoprecipitation assay, which was developed in the 1970s, was the first using snake toxins for labeling a neurological receptor. It has served as the prototype for the development of a wide variety of assays for anti-neurological receptor antibodies using radioactively labeled snake toxins which specifically bind receptors.

In the case of anti-AChR receptors in myasthenia gravis, bungarotoxin has an extremely low affinity constant for AChR, which means that the binding of bungarotoxin to AChR is so strong that the conditions of the assays will not separate them and thus the radioactively tagged bungarotoxin essentially strongly labels AChR. Once the AChR is labeled, the patient's serum is added to the muscle homogenates and anti-AChR antibodies bind to the labeled AChR. The next step is the addition of an anti-human IgG antibody which can precipitate antibody, and preferentially that antibody which is bound to a large molecule-like AChR. The precipitated material is washed free of unbound radioactive bungarotoxin and the remaining radioactivity measured in the precipitate is proportional to the levels of anti-AChR antibodies. The test is highly specific for myasthenia gravis, but a substantial proportion of MG patients, 15–50% depending on stage of the disease, are seronegative, i.e., negative on this test for anti-AChR antibody.

(iii) *Neuronal Antigens, Including Ion Channels and Receptors, in Paraneoplastic Neurological Disorders (PNDs).* A full discussion of assays for paraneoplastic antibodies is beyond the scope of this chapter. A recent review of paraneoplastic neurological disorders (PNDs) [4] identifies 17 different autoantibodies. The oldest and most well-studied autoantibodies in PND are anti-Yo (see clinical vignette of paraneoplastic cerebellar degeneration in Chap. 14) and anti-Hu antibodies. These bind to antigens in cerebral neurons and Purkinje cells (see Chap. 2). Initially, the reactivities were found by reacting serum from patients with PNDs with tissue slices, similar to the screening tests which detected anti-aquaporin 4 antibodies in NMO (see above). Many of these assays are now performed with purified antigens rather than whole tissue.

For testing for some of the anti-channel or anti-receptor antibodies, animal toxins are used to label the receptor protein prior to an immunoprecipitation assay in a methodology similar to the anti-acetylcholine receptor assay. The two most commonly used radioactive labeled toxins are the mamba toxin

dendrotoxin which labels voltage-gated potassium channels (VGKCs) in neuromyotonia and limbic encephalitis, and the snail toxin conotoxin which labels voltage-gated calcium channels (VGCCs) in Lambert-Eaton myasthenic syndrome (LEMS).

Another approach to testing for anti-receptor or anti-channel antibodies is to express the receptor in a cell line such as oocytes or fibroblasts and then test for binding of immunoglobulin from patient's serum to the cells. This has proven quite successful for anti-aquaporin receptor antibodies in NMO and has provided for more reliability than the immunofluorescence assays with whole brain tissue.

2.3   Blood Tests for Detecting Specific Pathogens

2.3.1   Cultures

The standard technique for identifying an infection in any body fluid is by sampling the body fluid, injecting the sample into culture medium, and identifying the responsible bacterium in the culture medium after it has replicated. Both bacteria and fungi can generally be readily identified in this way.

2.3.2   Polymerase Chain Reaction (PCR)

PCR takes advantage of the ability of DNA polymerase to copy single-stranded DNA, and uses primers, short complementary sequences of DNA, to identify the particular segment of the DNA which will be copied. What is thus needed for the test to work appropriately is correct sequences of the organism and intact DNA from the organism; if the organism is an RNA virus, it can be converted into DNA by using the enzyme reverse transcriptase; the PCR for RNA fragments then is called an RT-PCR. The assay is usually quite sensitive and specific. The PCR technique was invented and initially developed by Kary Mullis (see Inset 18.2).

The diagnostic PCR most commonly used in neuroimmunology is the detection of HIV in the

Kary Mullis received the Nobel Prize in 1993 for his discovery of the PCR, which revolutionized biology. Mullis' career has epitomized the ups and downs of many scientists who can work for decades on projects which can eventually be dead ends, or be huge successes which change the nature of science. After his graduate education Mullis left science to be a fiction writer, but later returned to science working a medical school. He then become a baker for a few years, but ultimately got a job with Cetus Corporation in California as a chemist. The idea for PCR came to him while he was driving one night, and Cetus allowed him time to pursue his idea. After some disappointing trials, the idea eventually bore fruit, and he received a $10,000 bonus from Cetus. Mullis has had some unusual ideas since then, discussed in his 1998 autobiography, Dancing Naked in the Mine Field, including a business selling jewelry containing amplified DNA of celebrities and an encounter with an extraterrestrial in the form of a fluorescent raccoon.

blood. This test both directly tests for the presence of the virus, and, if the test is positive, also quantitates the concentration of the virus in the blood. The ease with which HIV can be quantitated in blood revolutionized the field of HIV therapeutics since it allowed the rapid determination of viral load, which could be readily targeted by candidate therapies.

3   Cerebrospinal Fluid (CSF)

CSF is usually obtained via a lumbar puncture. Assays of the CSF are very helpful in neuroimmunology since the CSF circulates throughout the CNS and around large areas of the PNS. CSF analysis is generally a necessary part of the diagnostic workup in patients with neuroimmunological disease.

There are a few potential problems with the use of CSF for diagnosis. First is the potential false positivity in serological assays caused by high levels of immunoglobulin. CSF generally has very low immunoglobulin levels because of the blood–CSF barrier, but in a number of inflammatory conditions the immunoglobulin levels can rise to high levels. Since cutoffs for positivity for reactivity on antibody assays are dependent on normal CSFs with low immunoglobulin levels, and nonspecific binding increases with higher immunoglobulin levels, low-level false-positive assays are frequently seen in infectious or inflammatory conditions (see Inset 18.3). Clinical vignette-Liver damage as a result of a medical error in interpreting CSF antibody assays. Secondly, it is not unusual to have a so-called "traumatic tap" when lumbar punctures are performed, resulting in blood vessels being nicked in the process of inserting the needle. This results in some blood leaking into the CSF samples. Unfortunately, the presence of blood contamination in the CSF lessens the utility of the CSF as a diagnostic tool, and all possible steps should be taken to ensure that the spinal tap is performed cleanly. An example of an assay that is made inaccurate by a contaminated tap is one which tests for an "antibody index" in which levels of antibody levels in the CSF than the blood are measured relative to concentration of Ig in the blood, with the control for BBB permeability being the albumin in the two compartments. These assays are almost impossible to interpret with blood contamination of the CSF since the normal amount of IgG in the blood is 300–1,000 times that of the CSF. Thus, the presence of even a small amount of blood in the CSF will change the index.

A 35-year-old woman was referred to our medical center in 2007 for management of "severe treatment-refractory Lyme neuroborreliosis" to determine whether "continued antibiotic therapy was necessary given the risks."

(continued)

Inset 18.3 (continued)

She well until 2004 when she developed left optic neuritis which resolved over a few months with no therapy. In March 2006, she developed numbness in her left arm which resolved in a few weeks with no therapy. In 2007, she developed some ataxia and was evaluated by a neurologist who felt that she had Lyme disease based on a positive anti-*B. burgdorferi* CSF antibody assay. He referred her for treatment to an infectious disease specialist who treated her with ceftriaxone 1 g twice per day for 5 months, during which she developed abdominal pain, and was found to have severe cholecystitis and moderate hepatitis both felt to be secondary to the ceftriaxone. There was no history of rash, arthritis, facial palsy, or symptoms of meningitis.

The CSF results were reviewed and, in addition to showing a low-positive anti-*B. burgdorferi* CSF antibody assay was positive for OCBs. IgG index was not sent.

On exam, she had mild optic atrophy, diffuse hyper-reflexia, a positive Lhermitte's sign (flexion of the neck causing tingling sensations into the arms or legs), and mild cerebellar ataxia. An MRI of the brain showed multifocal white matter disease consistent with multiple sclerosis; one white matter lesion enhanced with contrast. Blood serologies for anti-*B. burgdorferi* antibodies were negative.

A lumbar puncture was repeated, showing once again oligoclonal IgG bands and a low-positive anti-*B. burgdorferi* antibody. However, the IgG index was exceedingly high with a very high CSF IgG. The anti-*B. burgdorferi* antibody index was negative indicating that the "positive" CSF anti-*B. burgdorferi* was a false positive, because of the very high CSF IgG concentration.

A diagnosis of multiple sclerosis was made, ceftriaxone therapy was terminated, her abdominal pain stopped, and her liver function tests slowly returned to baseline. Her subsequent clinical course was consistent with MS, and interferon-β therapy resulted in no adverse effects.

Author's note. Every laboratory assay must be assessed for its sensitivity, specificity, and predictive value in the population in which testing occurs. A positive test at a low level for anti-*B. burgdorferi* antibody in the CSF has a high sensitivity rate, but a very low specificity, and very low predictive value in patients who have high-IgG levels in the CSF. This is because immunoglobulin molecules are "sticky," i.e., like many proteins they have multiple interactions with other molecules which result in some nonspecific binding. Binding of one molecule to another can be measured by dissociation constants, which measure how strongly one molecule binds to another. This measurement is affected by such forces as electrostatic interactions, hydrogen bonding, hydrophobic, and van der Waals forces. Thus, for any particular binding of an antibody to an antigen, some will be nonspecific binding and some will be specific binding. Affinity is the inverse of dissociation so an affinity constant for an antibody binding to its antigen is the inverse of the dissociation constant. As the concentration of an immunoglobulin in a particular solution increases, it will be more likely to bind to other molecules nonspecifically. Thus, most binding assays are highly concentration dependent. For blood tests, this is not usually an issue since the immunoglobulin concentration in the blood of most humans is highly regulated and is mostly the same from one individual to another. However, CSF immunoglobulin concentrations is highly variable from one disease process to the next and can be much higher in patients with MS or other CNS inflammatory processes than in normals. Thus, assays which measure binding of immunoglobulin in the CSF to the spirochete *B. burgdorferi*, as well as other antigens, will frequently be false positive at a low level in patients with MS. This is the reason antibody indexes are helpful, because binding is corrected for the relative concentrations of immunoglobulin in the fluids.

3.1 Antibodies in the CSF

3.1.1 General Approach to Determination of Intrathecal Antibody Production of Any Specificity

The diagnosis of many neuroimmunological diseases is aided by measuring immunoglobulin populations within the CSF. There are two broadly defined types of assays: *nonspecific*, measuring IgG generally, or *specific*, which measure antibodies of a particular antigenic specificity. In both, the contribution of immunoglobulin from the blood needs to be determined, since there is an equilibrium of CSF IgG with blood IgG.

QIgG, Qalbumin, and the IgG Index

In the normally functioning CNS, the ratio of IgG concentration in the lumbar CSF relative to that of blood is about 1/300 to 1/1,000; this ratio is referred to as Q_{IgG}. In normal circumstances, all of this IgG is derived from IgG in the blood, and none is produced within the CNS, since there are no plasma cells producing IgG within the CNS unless there is an inflammatory CNS process. When the blood–CSF barrier is disrupted, such as in acute meningitis, the IgG concentration in the CSF increases as more IgG from the blood is able to pass into the CSF. In contrast to meningitides, not all CNS inflammatory processes result in significant disruptions in the blood–CSF barrier; e.g., in multiple sclerosis there is usually not adequate diffuse change in blood–CSF barrier to markedly change immunoglobulin concentrations and most of the IgG present in the CSF in MS is produced by B cells within the brain. Thus, in an inflammatory process, it may be important to determine whether the elevated IgG in the CSF is produced within the CNS, or simply as a consequence of an impaired blood–CSF barrier. This can be done by comparing the IgG ratio, i.e. Q_{IgG}, to a ratio of albumin in the CSF relative to the serum, called $Q_{albumin}$. Albumin is another large protein found at a 300–1,000-fold concentration higher in the serum than the CSF. The IgG index is derived from Q_{IgG} divided by $Q_{albumin}$. Thus, if there is a production of IgG within the CNS, the IgG index will be higher than 1. Graphs plotting

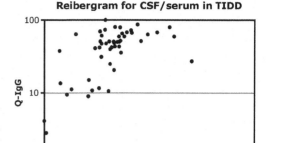

Fig. 18.2 A Reibergram demonstrating intrathecal IgG production in a mouse model of MS [9]. In this Reibergram [5], CSF/serum indices for IgG and albumin are displayed on a log–log scale, demonstrating intrathecal synthesis of IgG. The x-axis is the Q-albumin, or albumin index (CSF albumin/serum albumin) in which the upper limit of normal is 8. The y-axis is the Q-IgG, or IgG index (CSF IgG/serum IgG), in which the upper limit of normal is 3. Each point is the CSF analysis of a single mouse with Theiler's virus-induced demyelinating disease. The albumin and IgG indices are multiplied by 1,000 for clarity; IgG in the CSF is approximately 1/1,000 that of the serum

these relationships are sometimes called "Reibergrams" after Hansotto Reiber, a German investigator who has popularized the application of these analyses to neuroinflammatory diseases [5] (see Fig. 18.2). The Reibergram can provide more information to a clinician than the IgG index, because the IgG index does not provide information about the integrity of the blood–CSF barrier.

Assaying CSF for Oligoclonal Bands

Assaying CSF for oligoclonal bands (OCBs) is a test of the CSF in which proteins are separated by an electrophoretic technique and then identified as IgG usually by an anti-IgG antibody. Similar IgGs group together and a specific immune response is characterized by such aggregation, so that an immune response in which plasma cells in the brain produce antibody can be identified by positive OCBs in the CSF. Since some OCBs can "leak" into the CSF from the blood, a truly positive OCB analysis is one in which OCBs are detected in the CSF which are not detected in blood. The test is particularly helpful in multiple sclerosis in which OCBs specific for the CSF using optimal techniques are present in over 90% of patients with MS, frequently when no other immune test is abnormal in CSF or blood. This figure of over 90% sensitivity for OCBs in MS comes from studies in which there is careful attention to details of the OCB assay. In many practice settings in the USA, however, CSF analysis is not performed optimally, and the predictive value of the assay will subsequently be lower. In a paper presented at the Consortium of MS Centers meeting in 2010, Rauchway et al. conducted a survey of 225 laboratories throughout the USA and found that only 61(27%) were performing OCB testing according to consensus conference recommendations [6]. Physicians caring for MS patients should be cognizant of the quality of the assays in the CSF available to them, since, with acceptable CSF OCB assays performed for the diagnosis of MS, a "negative assay strongly suggests an alternative diagnosis" [6].

3.1.2 Antigen-Specific Tests of the CSF
Reaginic Tests for Syphilis
Neurological involvement in syphilis can be difficult to detect, and a positive CSF reaginic antibody (see reaginic antibody above) determines that *Treponema pallidum* has invaded the CNS and resulted in plasma cells within the CNS producing reaginic antibodies. Thus, patients with neurosyphilis will have a positive CSF VDRL or CSF RPR; sometimes reaginic tests can be positive in the CSF when they are not positive in the blood.

Lyme Disease
Infection with *Borrelia burgdorferi* resulting in Lyme neuroborreliosis is an increasingly common infection in many parts of the world, and cannot be readily detected by assays testing for the presence of the pathogen. An antibody index in which a ratio of the level of anti-*B. burgdorferi* IgG antibody in the CSF relative to that of serum is determined and compared to the relative amounts of total IgG is a very helpful assay for the detection of Lyme neuroborreliosis. This antibody index is the test of choice for Lyme

neuroborreliosis. The absence of significant anti-*B. burgdorferi* antibody in the CSF is evidence against the diagnosis of Lyme neuroborreliosis.

Viral Antibodies
Prior to the development of PCR to detect viruses, antibody assays were frequently the only way to detect the presence of the viruses in the CNS. However, this approach has many false negatives, and often by the time viral antibodies are detected in the CSF, the pathogen has been cleared, or has caused major injury. In contrast, since viruses usually have a large copy number relative to other types of infections and PCR is a sensitive assay for the nucleic acid of a virus, PCR has become test of choice in the CSF for detecting viral infections in the CNS. CSF viral antibodies are now rarely used.

3.2 Detection of Pathogens in the CSF

3.2.1 Polymerase Chain Reaction in the CSF
Both for PCR and culture, the accuracy of the test is decreased if blood leaks into the CSF during the LP. Thus, obtaining an "atraumatic" spinal tap is highly desirable. Some tissue proteins can interfere with function of the DNA polymerase, but the CSF is usually a remarkably "clean" body fluid in this regard and excellent sensitivity is obtained from CSF PCRs, as long as significant amounts of blood do not contaminate the CSF. If there are very few copies of the pathogen, or if the organism is an RNA virus that requires reverse transcriptase to convert the RNA to DNA, sensitivity may be less than optimal. Thus, *Borrelia burgdorferi* generally has a very low copy number in the CSF resulting in relatively low sensitivity of CSF PCR in Lyme neuroborreliosis.

3.2.2 Culture of the CSF
The clinical situation usually dictates what type of cultures is ordered. A patient presenting with a hyperacute meningitis, with a CSF showing large

numbers of polymorphonuclear leukocytes and a positive gram stain consistent with a pyogenic meningitis, requires bacterial cultures. In a patient with a subacute or chronic meningitis, and CSF mononuclear pleiocytosis, the CSF should be sent for fungal cultures as well as a cryptococcal antigen determination by latex agglutination. In addition, it is helpful to obtain an extra tube of a minimum of 2 cm^3 CSF for short-term storage in a refrigerator, because frequently more testing is indicated after the first round of results are obtained. Having an extra tube allows more directed tests to be performed without the need for a repeat LP; for most types of cultures, an overnight incubation at 4°C will not significantly decrease the culture yield. Thus, the morning after the initial testing of the CSF, more CSF could be sent for other organisms if within the differential.

3.2.3 Viruses

The diagnosis of viral infections of the CNS has been revolutionized by the availability of PCR to detect viral infections. In the past, the diagnosis of these infections was frequently very difficult, but now the sensitivity and specificity of viral detection in the CSF is so high that many neuroimmunologists are hesitant to make the diagnosis of viral infections such as herpes simplex encephalitis, varicella zoster-associated vasculopathies, or PML without positive CSF PCRs for the virus. However, unlike culture techniques, in which one can simply order "bacterial culture" or "fungal culture," in PCR, specific sequences of a virus are utilized so one must request a PCR of the specific pathogens which are high on the list of likely organisms.

3.2.4 Tuberculosis

Mycobacterium tuberculosis is the most common cause of difficult-to-diagnose chronic meningitis, and PCR has proven very helpful in its diagnosis. It should be strongly suspected in patients who have meningitis with more than 5 days of symptoms, and less than 1,000 white cells per cu.mm. in the CSF of which lymphocytes form the majority [7].

3.2.5 Fungi

Cryptococcus, coccidioides, aspergilla, and histoplasma are not rare causes of chronic meningitis (see Inset 12.3 for coccidioidal meningitis); CSF PCR can be helpful for their diagnosis, although culture is often more easily available.

3.2.6 Other Pathogens

Many other infections in the CNS are not associated with spillage of pathogen into the CSF that is adequate to allow detection of the organism by PCR. This is especially true if meningitis is not a prominent part of the infection, and the organism is contained in the parenchyma of the brain. Walling off of the infection within the parenchyma—e.g., abscess formation—reduces ascertainment even further. Examples of organisms implicated in such situations include toxoplasma and *Taenia solium* (the cause of cysticercosis).

4   Tests of Immune Function

Increasingly patients with neuroimmunological diseases are being treated with immunosuppressive medications as reviewed in Chap. 19. Monitoring for immune function is in its infancy. It is likely that there will be significant advances in this area as immunosuppressives are increasingly used for chronic diseases.

4.1   Humoral Immunity

Total immunoglobulin levels in the serum or measurement of specific isotypes can be helpful to detect potentially dangerous impairments of humoral immunity. This can be especially important in therapies targeting B cells. In a 1-year phase 2 study of the anti-CD20 therapy, rituximab, in MS, a single course of therapy resulted in significant incidence of lowered immunoglobulin levels, particularly IgM [8]. Some investigators are recommending that antibody responses to seasonal influenza vaccines be used as a measure of suppression of humoral immunity in immunosuppressed patients, but this interesting approach has not been tested.

4.2   Cellular Immunity

4.2.1   PPD

In tuberculosis, a sensitive test for the presence of the infection is demonstration of delayed-type hypersensitivity (DTH) to the bacteria by a skin test, frequently called a purified protein derivative (PPD). Small amounts of mycobacterial antigens are injected into the skin and within a few hours after injection, T cells and other inflammatory cells enter the area leading to local swelling and redness which can be measured as a thickened red area on the forearm if the individual had anti-mycobacterial immunity; this peak amount of inflammation is usually at about 2 days after injection. Individuals who did not have infection or exposure would not have swelling. A cause of a false-negative reaction is anergy, the inability to react usually because of immunosuppression. Anergy can be ruled out by injecting with a panel of antigens derived from minor infections to which individuals are commonly exposed, such as mumps, trichophyton, or candida; anergy is ruled out when a positive test develops. Thus, in most ill patients, a PPD should only be considered negative if there is in addition a positive response in one of the anergy controls. Increasingly, clinicians are using in vitro correlates of the PPD, such as the Quantiferon assay, in which lymphocytes from the blood are exposed to Mycobacterial antigens which produce activation of the lymphocytes in TB patients.

4.2.2   PHA stimulation tests

The functional capacity of human lymphocytes to rapidly divide upon stimulation can be tested by adding lectins to the cells and measuring their proliferation. Lectins are proteins that bind sugar molecules with high affinity; the binding of a lectin, such as phytohemagglutin (PHA) or concanavalin A(conA) to sugar molecules on the surface of lymphocytes activates them into DNA synthesis. These lectins are also sometimes referred to as mitogens. In the recent phase 2 study of the anti-CD25 monoclonal antibody, daclizumab, in MS, a PHA stimulation test called ImmuKnow was performed to evaluate immune system functionality. This test, marketed by Cylex, utilizes stimulation ex vivo by PHA of CD4-positive T cells in fresh whole blood, and ATP production is assessed. PHA stimulation tests like the ImmuKnow may be used in the future in neuroimmunological patients in whom excess immunosuppression may come with higher risks.

References

1. Gorelik L, Lerner M, Bixler S, et al. Anti-JC virus antibodies: implications for PML risk stratification. Ann Neurol. 2010;68(3):295–303.
2. Bradl M, Misu T, Takahashi T, et al. Neuromyelitis optica: pathogenicity of patient immunoglobulin in vivo. Ann Neurol. 2009;66(5):630–43.
3. Lennon VA, Wingerchuk DM, Kryzer TJ, et al. A serum autoantibody marker of neuromyelitis optica: distinction from multiple sclerosis. Lancet. 2004; 364(9451):2106–12. Dec 11–17.
4. Dalmau J, Rosenfeld MR. Paraneoplastic syndromes of the CNS. Lancet Neurol. 2008;7(4):327–40.
5. Reiber H. Cerebrospinal fluid–physiology, analysis and interpretation of protein patterns for diagnosis of neurological diseases. Mult Scler. 1998;4(3):99–107.
6. Freedman MS, Thompson EJ, Deisenhammer F, et al. Recommended standard of cerebrospinal fluid analysis in the diagnosis of multiple sclerosis: a consensus statement. Arch Neurol. 2005;62(6):865–70.
7. Thwaites GE, Schoeman JF. Update on tuberculosis of the central nervous system: pathogenesis, diagnosis, and treatment. Clin Chest Med. 2009;30(4):745–54. ix.
8. Hauser SL, Waubant E, Arnold DL, et al. B-cell depletion with rituximab in relapsing-remitting multiple sclerosis. N Engl J Med. 2008;358(7):676–88.
9. Pachner AR, Li L, Lagunoff D. Plasma cells in the central nervous system in the Theiler's virus model of multiple sclerosis. J Neuroimmunol. 2011;232:35–40.

Therapy in Neuroimmunological Disease

19

We practice neurology in a fog...We're ignorant. We're making decisions based on what our sense of best practice medicine should be, but we have no solid knowledge that that's actually the right thing.

S. Claiborne Johnston, M.D., Ph.D., executive vice editor
of the Annals of Neurology, from Neurology Reviews, June 2010

1 Introduction

The sentiments of Dr. Johnston's quotation about the clinical practice of neurology also apply to the more limited field of neuroimmunological therapies. Most of the situations in which neuroimmunologists prescribe immune suppressive therapies have not been subjected to randomized clinical trials (RCTs) (see Chap. 7) so such treatment is frequently on shaky evidentiary grounds for justification of its use. At times, when RCTs have been performed, the results have demonstrated that practices firmly entrenched in clinical practice have no merit. For instance, in the Optic Neuritis Treatment Trial, described in Chap. 7, the common practice of administering oral corticosteroids for optic neuritis was shown to be not helpful in that the rate of new episodes of optic neuritis in either eye was as high or higher in those receiving oral corticosteroids than in those receiving intravenous corticosteroids or placebo [1]. In many instances, neuroimmunologists will try to adhere to the physician's law, *primum non nocere*, i.e., the first rule is to do no harm, but the trend over time has been to be more aggressive and less conservative. Increasingly, neurologists and statisticians have worked together to attempt to define concepts such as "number needed to treat" (NNT) and "number needed to harm" (NNH) so that clinicians could have a better idea about risk/benefit ratios for drugs. Unfortunately, these types of analyses often provide information that is not highly supportive of drug use, and thus pharmaceutical companies are frequently loathe to do such analyses of their products in pharmaceutically company-sponsored clinical trials.

Most of the medications used by neuroimmunologists are meant to decrease inflammation or suppress immune responses. These medications are frequently called either immunosuppressive or immunomodulatory drugs; sometimes, the former term is used if the drug increases the risk of infection or cancer, while the latter is used if it does not have those unwanted effects. Both the power and adverse effects of most of these drugs are dose-dependent, with higher doses having more effect

A.R. Pachner, *A Primer of Neuroimmunological Disease*,
DOI 10.1007/978-1-4614-2188-7_19, © Springer Science+Business Media, LLC 2012

and also more likelihood of side effects. Thus, the terms immunosuppression and immunomodulation are often used interchangeably. Most of the drugs used have broad-spectrum effects on the immune system, while recently more specifically targeted drugs have been developed, primarily using monoclonal antibody technology.

Neuroimmunologists using these drugs must monitor patients for both their efficacy and their side effects. Monitoring for efficacy is usually done by analyzing the clinical effects of the drugs on the disease being treated, rather than the degree of effect on the immune system. Ideally, we could also monitor the effects of these drugs on the immune system, but there is an unfortunate lack of reliable measures of effect of these drugs on the immune system, i.e., no reliable biomarkers. This is a major problem in the field, since there is a broad range of response seen in individual patients, and one is never sure that an adequate degree of immunosuppression is being reached. It is especially a problem if the disease is not responding well to the treatment. In such situations, the neuroimmunologist does not know whether the illness is simply severe and thus refractory to a normal level of immunosuppression, or alternatively, if the patient is not adequately immunosuppressed. Different neuroimmunologists have their own ways of dealing with this conundrum. Many simply raise the dose of the medication to the point of toxicity; others stop the ineffective drug and try others.

With respect to side effects of medications, neuroimmunologists differ in their philosophy. Some are very conservative, adhering to the Hippocratic dictum "primum non nocere" (the first rule is to do no harm), and reserving potentially dangerous therapies for only the most severely involved patients. Others are more aggressive attempting to eradicate even the last vestiges of an illness, accepting an occasional severe adverse outcome as an acceptable risk. Most neuroimmunologists function between these two extremes.

In this chapter, neuroimmunological therapies used in MS and other neuroimmunological diseases are reviewed; more information on therapies used primarily in MS can be found in Chap. 7.

2   Chemicals

2.1   Corticosteroids

Corticosteroids are by far the most widely used immunomodulatory drugs in the management of neuroimmunological disease. They are effective in a wide variety of neuroimmunological diseases, including multiple sclerosis, myasthenia gravis, chronic inflammatory demyelinating polyneuritis, and inflammatory muscle disease; they are also used in suppressing inflammation in infections such as tuberculous meningitis. They are a group of chemically related molecules that resemble cholesterol in structure, and endogenous versions are produced in the adrenal cortex. They are usually taken by mouth, but intravenous preparations are available when higher doses or more rapid effects are needed. They are thought to function through effects on gene expression by directly affecting transcription factors within the nucleus, one of the most prominent of which is nuclear factor-kappa B (NF-kappa B) (Inset 19.1).

Inset 19.1 Nuclear Factor Kappa-Light-Chain-Enhancer of Activated B Cells (NF-Kappa B)

NF-kappa B is a complex of proteins which, upon activation, can act relatively quickly in the nucleus to upregulate expression of a variety of genes involved in proliferation and cell survival. In the absence of activation stimuli, NF-kappa B components remain in the cytoplasm because of their interaction with inhibitors of NF-kappa B (I-kappa Bs). NF-kappa B is considered a rapidly acting transcription factor. A prominent class of molecules which participate in its activation is toll-like receptors (TLRs). Corticosteroids function in part by interferring with NF-kappa B, while pathogens upregulate NF-kappa B, frequently through TLRs. There is evidence, for instance, that *Borrelia burgdorferi*, the causative agent of Lyme neuroborreliosis upregulates NF-kappa B via TLR2.

2.1.1 Mechanism

The effects of pharmacologic corticosteroids can be divided into those which have primarily a metabolic action (mineralocorticoids) and those with predominantly immune and inflammatory-related mechanisms of action (adrenocorticoids). Commercially available steroid therapies are usually some mix of these two characteristics of these two actions. They cause decreased numbers of lymphocytes in the peripheral blood, but increased numbers of neutrophils. Cytokine secretion by T cells is decreased and B-cell function is impaired, particularly development of antibody to new antigens. Many components of the innate immune response are also impaired.

2.1.2 Side Effects

Side effects of short-term steroid dose are dose-related, and if low doses are used, adverse effects are minimal. Short-term, high dose use can cause problems, including elevated blood glucose, especially in patients at risk for, or already diagnosed with diabetes, acneiform skin disease, as well as changes in mood, personality, and problems with sleep.

Long-term steroid use is limited by a much longer list of potential serious side effects. Along with chronicity of the above problems, the most serious problems include, but are not limited to:

- Bone problems including avascular necrosis, osteoporosis with fractures.
- Body fluid problems, including edema, weight gain, and sodium retention.
- Connective tissue disorders, including impaired wound healing, thinning, and easily cut skin.
- Muscle weakness because of damage to muscle cells.
- Ophthalmologic problems, including cataracts.
- Inflammation of the pancreas.
- Masking of the signs of infection; this problem is especially serious because sometimes infections can be life-threatening but elicit minimal signs because of the potent anti-inflammatory actions of corticosteroids (see Inset 19.2).
- Sleep disorders.
- Psychiatric side effects mostly agitation and anxiety.

Inset 19.2 A College Student on Corticosteroids with Malaise and Fatigue

J.L. was a 24-year-old graduate student who was admitted to the hospital after a fall because of bilateral leg weakness. She had systemic lupus erythematosus (SLE), an inflammatory systemic disease affecting multiple organs, but had been doing very well until 4 months prior to her admission when she had a worsening which required moderate daily corticosteroid therapy. Her internist had considered lowering her dose a month previous to her hospitalization, but had decided not to. She had returned to her internist 1 week prior to hospitalization with complaints of "not feeling well" and "tired throughout the day," but examination of the patient and routine blood work had been unchanged from previous evaluations. On the day of her hospitalization, she had felt worse, and when walking away from the table at dinner had fallen and had trouble getting up. She called an ambulance and was brought to the ER. By the time, she had received routine evaluation and blood work, it was late at night. The blood work showed only a slightly higher neutrophil count than her normal which always was elevated when she was treated with corticosteroids. She continued to have trouble walking in the ER because of bilateral leg weakness. The neurologist was concerned about the bilateral leg weakness but could find no other worrisome signs and thought that this might be steroid-induced muscle weakness. He ordered an MRI of the cervical and thoracic spine, but more severe cases were ahead of her and because of the lateness of the hour, the patient was admitted to the neurology ward, with imaging scheduled for the next day. However, the evening of admission, she began to develop severe back pain, a fever, and worsening leg weakness, and quickly progressed to confusion and hypotension. Blood cultures were obtained, antibiotics

(continued)

were begun, and emergency imaging of her spine showed a large epidural abscess. She rapidly deteriorated; emergency surgical consult was called, but she was considered too precarious to undergo anesthesia, and she died of sepsis in the morning. Blood cultures grew *Staphylococcus aureus*.

Author's note. Spinal epidural abscess is almost never an easy diagnosis to make but usually back pain, slowly progressive weakness, and an abnormal white count over a number of days leads to the correct diagnosis. In this patient, the corticosteroid therapy blunted the symptoms of back pain and the development of the weakness; the white count was not clearly abnormal because corticosteroids normally elevate the neutrophil count. This patient's story demonstrates masking of symptoms of infection, one of the many dangerous side effects of chronic corticosteroid use.

Because of the extensive potential side effects of long-term therapy, the benefit/risk ratio for using these drugs must constantly be addressed during therapy. This assessment is made even more difficult by the unpredictability of many of these side effects. These side effects are also to some degree dose-related so that the clinician must always attempt to use the minimum dose possible.

These side effects of corticosteroids are most likely with higher doses and more prolonged therapy. Thus, most of the time corticosteroids are administered for relatively short periods, a few days to at most a few weeks, and then tapered down to discontinuation or to alternate day therapy. Alternatively, especially if there is no emergent need to induce a remission quickly, alternate day steroids can be initiated at low doses with subsequent increase in dose as needed [2].

2.1.3 Monitoring

Careful monitoring for both optimal benefit and minimum adverse side effects is necessary, especially for long-term use.

(a) Monitoring for optimal benefit: Corticosteroids predictably increase the peripheral white cell count in most patients, which frequently causes consternation about elevated white cell counts, since raised white cells in the blood is often a harbinger of underlying infection, and infections are more common in corticosteroid-treated individuals. However, raised white counts are not indicators of optimal treatment effect, since good outcomes can be achieved when white counts are normal, and inadequate responses can be seen with high white counts. There are no good biomarkers of corticosteroid effect, so neuroimmunologists will carefully assess how well the corticosteroids are downregulating the inflammatory and immune aspects of the disease process and adjust steroid doses accordingly.

(b) Monitoring for adverse effects: All of the above potential side effects of chronic corticosteroid therapy need to be monitored. Patients with diabetes mellitus must be very carefully watched using assessments of blood sugar and hemoglobin A1c, a glycosylated hemoglobin which increases with poor control of diabetes over time. Even patients without diabetes can become hyperglycemic, sometimes dangerously so, when they are treated with corticosteroids, especially if there is diabetes in the family. Psychiatric problems such as depression or mania are common and can require medication for control. Bone density measurements at baseline and at intervals during treatment are useful. Body weight needs to be monitored as excessive salt and water retention sometimes can require diuretic therapy.

2.2 Immunosuppressives

2.2.1 Azathioprine (Imuran in the USA)

A small chemical (molecular weight of 277), azathioprine is one of the original immunosuppressive medications, first used in the 1960s in organ transplantation and in MS not long thereafter. It is still used extensively in transplantation, although it is being replaced to some extent by mycophenolate (see below). It is a purine analog

Fig. 19.1 Thiopurine drug metabolism. The figure is a representation of thiopurine drug biotransformation, with azathioprine being converted in vivo to 6-mercaptopurine (6-MP), followed by the metabolic activation of 6-MP to form 6-thioguanine nucleotides. 6-MP undergoes metabolism catalyzed by xanthine oxidase (XO) or TPMT

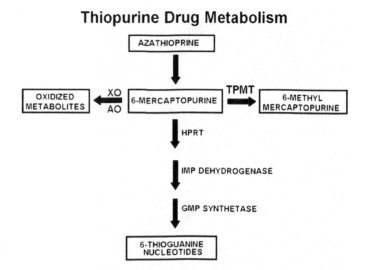

and a DNA synthesis inhibitor and is metabolized to 6-mercaptopurine and 6-thioguanine and 2 inactive metabolites (see Fig. 19.1). Thus, rapidly dividing cells, such as lymphocytes during an active immune response, are inhibited. Administered as a once or twice daily oral dose, it is very popular for MS in Europe, listed as a "basic" drug for MS, along with IFN-β and GA, by the Multiple Sclerosis Therapy Consensus Group in 2008, consisting of 109 European MS experts [3]. A recent study in Italy demonstrated that azathioprine was very effective in decreasing gadolinium-enhancing lesions in MS [4]. It is used much less frequently for MS in the USA, relative to Europe.

Another use of azathioprine in neuroimmunological diseases is as a "steroid-sparing" agent. Since chronic high-dose corticosteroid therapy may have a high incidence of unacceptable side effects, patients who require prolonged therapy, such as those with inflammatory neuropathies, myopathies, or myasthenia gravis, will frequently be treated with azathioprine in conjunction with lowering the dose of corticosteroids, ideally to the point of using alternate-day corticosteroids.

The most severe potential risks of azathioprine are carcinogenesis and teratogenesis, although the incidence of both of these seems to be extremely low in MS patients when other immunosuppressives are not used concurrently. The risk of teratogenesis is so low that, when used to immunosuppress for transplant rejection, discontinuation is not recommended when the patient becomes pregnant. For all of these reasons, azathioprine's relative popularity is due to its generally being considered the safest and easiest of the immunosuppressives.

2.2.2 Methotrexate

Methotrexate acts on rapidly dividing cells by being a folic acid antagonist, resulting in decreased nucleotide synthesis. It is one of the oldest immunosupressives currently used, having been developed, initially as a cancer chemotherapeutic agent, by Yellapragada Subbarao at Lederle Laboratories in the late 1940s. The most common use in the USA is in cancer chemotherapy and, at much lower doses, rheumatoid arthritis (RA). For most indications, methotrexate can be administered as a once weekly oral dose. In RA, it is the most popular of the immunosuppressives for the following reasons: it is effective, patients are adherent with the medication, other agents can be added to it, side effects are uncommon, and those side effects which do occur (hepatitis or bone marrow suppression) can be easily monitored with routine blood tests. It is only rarely used in MS by American neurologists, and, within neuroimmunological therapeutics, is more commonly given as a second-line or steroid-sparing agent for the inflammatory myopathies.

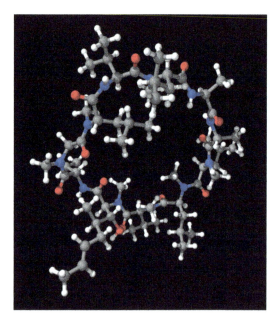

Fig. 19.2 Cyclosporine, with its distinctive cyclical structure, was isolated by growth of the fungus, *Tolypocladium inflatum*, from a soil sample in Norway in 1969, by Sandoz (now Novartis) scientists. Its structure is a cyclic peptide containing 11 amino acids. The drug was initially developed for its effect in the prevention of transplant rejection. Some of the first studies were in liver transplantation

2.2.3 Mycophenolate (CellCept or Myfortic)

This relatively new compound, a purine synthesis inhibitor derived from the fungus Penicillium, has been extensively used in the transplant population, including kidney transplantation, and has been increasingly tried in neuroimmunological diseases, such as MS and NMO [5]. Its use has lessened somewhat with reports of PML (see Chap. 12) in patients taking the drug, usually in transplant patients.

2.2.4 Cyclosporine

This drug is produced by the fungus *Tolypocladium inflatum* (see Fig. 19.2). One of its unique features is that it contains a D-amino acid, rarely encountered in nature, in contrast to the L optical isomeric form of amino acids normally found in proteins. The mechanism of action of cyclosporine is the inhibition of

cytokine release, primarily IL-2 by T cells. This happens via the binding of cyclosporin to cyclophilin and subsequent inhibition of the phosphatase calcineurin, an important molecule in IL-2 transcription. Related drugs, used primarily for organ transplantation and not in neuroimmunology, are sirolimus (rapamycin) and tacrolimus (FK506). Sometimes, these drugs, cyclosporin, rapamycin, and tacrolimus, are grouped together and referred to as calcineurin inhibitors (CNIs). Cyclosporine and the other CNIs have been extensively used in transplantation, and their introduction in that field has greatly cut down the rate of transplant rejections. However, CNIs can have similar side effects to those associated with corticosteroids and thus the use of CNIs with corticosteroids, with the attendant potential for cumulative adverse effects, needs to be especially closely monitored.

2.2.5 Cyclophosphamide (Cytoxan)

This drug is a small molecule (MW = 261) related to the nitrogen mustards developed for chemical warfare and is a nonspecific DNA alkylating agent. It suppresses cell replication and thus both cell-mediated and humoral immunity. It has been used for at least 50 years as a chemotherapeutic agent in a variety of malignancies but mostly for lymphomas. In contrast to azathioprine, cyclophosphamide is considered a stronger immunosuppressive agent and is generally felt to be more effective than less potent immunosuppressives in suppressing inflammation in patients with very active inflammatory disease. However, the flip side of that coin is that adverse effects of cyclophosphamide are also more likely (see Inset 19.3). Its immunosuppressive effects are dose-dependent, and at low doses, cyclophosphamide can be immunostimulatory, possibly because of preferential effect of low doses on regulatory T cells (see T_{reg} cells in Chap. 1). Some MS specialists use cyclophosphamide relatively frequently, finding it a helpful adjunct in patients not responding to safer medications [6], but most neurologists use cyclophosphamide for neuroimmunological diseases only rarely.

Fig. 19.3 Fingolimod's structure resembles that of sphingosine, a naturally occurring glycolipid which is an important molecule for lymphocyte trafficking through lymph nodes

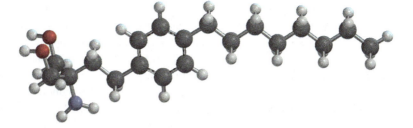

2.2.6 New Drugs

Fingolimod

This medication, whose brand name is Gilenya, is unique for a number of reasons. First, its structure is that of a sphingolipid (Fig. 19.3), indicating that is belongs to a large group of neurologically important, but poorly understood molecules, which include glycosphingolipids, such as gangliosides and cerebrosides. Second, it is the first of a class of drugs which focus on interactions between sphingolipids and their receptors. For fingolimod, the biologically relevant ligand and receptor are sphingosine and sphingosine-1 phosphate (S-1 P) receptor, which is a critically important receptor on many classes of lymphocytes, including T- and B cells, and especially important in lymphocyte trafficking [7]. The drug functions by causing internalization of S-1 P receptors leading to interference with normal ligand–receptor function. Fingolimod's effect of decreasing attacks in MS without inducing major immunosuppression with coincident opportunistic infections is thought to be due to its preferential retention of subpopulations in lymph nodes while affecting other subpopulations less [8]. Specifically, it is thought to retain naïve T cells and central memory T cells in lymph nodes while having less of an effect on peripheral effector memory T cells; it is the peripheral effector memory T cells which are thought to be most important during acute infections. The retention of lymphocytes in the lymph nodes leads to substantial lymphopenia in the peripheral blood, with total lymphocyte counts in treated patients generally being 20–30% of pretreatment values. This is an expected effect of the drug, and indeed can serve as a useful biomarker of compliance. Fingolimod also accumulates in the CNS [9], and neuroprotective effects are possible and hoped-for but not proven. Fingolimod was approved for the treatment of MS in September 2010, and since its approval and has been increasingly used as the first oral MS disease-modifying drug. It is the first sphingosine agonist to be used for the treatment of disease.

Cladribine

This medication is FDA approved for the treatment of hairy cell leukemia and is approved in other countries for the treatment of chronic lymphocytic leukemia. The mechanism of action is via disruption of cellular metabolism by an active metabolite of cladribine, 2-chlorodeoxyadenosine triphosphate. Lymphocytes are preferentially affected. Cladribine treatment results in sustained reductions of both CD4 and CD8 counts. Immunosuppression is the result of therapy, although opportunistic infections have not been a prominent feature of clinical trials of cladribine in MS. After a promising start and high hopes for approval for MS in the European Union and the USA, the drug was not approved by either, and the drug faces an uncertain future in MS globally.

3 Biologics: Large Molecules Produced via Recombinant Technology or from Human Specimens

3.1 Intravenous Immunoglobulin

Intravenous immunoglobulin (IVIg) is used extensively in neuroimmunological disease, and consists of human IgG purified from pooled blood from normal humans. The FDA requires that, prior to use in the USA, a given batch be an aggregate of blood obtained from at least 1,000 donors. IVIg was originally developed to treat children with Bruton's type agammaglobulinemia (see Chap. 1), a congenital condition due to a mutation in an enzyme, and other genetic conditions causing low serum immunoglobulin levels (hypogammaglobulinemias). IVIg was subsequently found to have immunomodulatory effects, and now is used to treat a broad range of immune-mediated conditions. Its mechanism of action is unknown and may be due to a suppression of complement activity, downregulation of autoantibody production, blockade of Fc receptors, or downregulation of inflammatory cytokine production. Its most appealing characteristic is safety, i.e., IVIg therapy does not result in increased susceptibility to infection or to increased risk of malignancy, and there are no other major systemic adverse effects such as kidney or liver injury. The main disadvantage is its cost, which, for chronic conditions, is often more than $50,000 USD per year.

IVIg can be considered effective in Guillain–Barré syndrome, chronic inflammatory demyelinating polyneuropathy and multifocal motor neuropathy, but its efficacy in other conditions is less well-documented. Some investigators feel it is a second-line agent for dermatomyositis and stiff-person syndrome. Despite lack of hard evidence for positive effect in most situations, it has been increasingly used in a variety of other neurological diseases, prompting concerns related to the increasing cost and lack of demonstrated efficacy [10]. For instance, it continues to be used in multiple sclerosis, despite the results of a recent large multi-center, randomized, double-blind, placebo-controlled study [11], which demonstrated no efficacy relative to placebo for the use of 0.2 or 0.4 g/kg every 4 weeks for 1 year.

3.2 Cytokines

At this point in time, the only recombinant form of a cytokine being used in neuroimmunological therapy is IFN-β. This molecule is the most commonly prescribed disease-modifying therapy for MS; its use in that disease is described in Chap. 7.

3.3   Monoclonal Antibodies (See Sect. 2.1 in Chap. 17)

Natalizumab was the first monoclonal antibody approved for a neurological disease when it was approved by the FDA in 2005 for the treatment of MS. Three other monoclonals, already approved by the FDA for the treatment of cancer or transplant rejection have had promising phase 2 studies in MS and are being evaluated in phase 3 studies.

(a) Monoclonal antibody to α-4 integrin (natalizumab, Tysabri): This biologic, reviewed in Chap. 7, is currently being prescribed for tens of thousands of patients worldwide.

(b) Monoclonal antibodies to CD20 (rituximab or ocrelizumab): These biologics deplete cells in the B-lymphocyte lineage by binding to the CD20 antigen on their surface and is discussed in Chap. 17. Rituximab is not FDA-approved for use in any neuroimmunological disease, including MS, but has been used for lymphoma since its approval for that cancer in 1997. Phase 3 trials in MS are now under way to determine efficacy, after demonstration of substantial efficacy in a phase 2 study [12].

(c) Monoclonal antibody to CD52 (alemtuzumab, Campath): CD52 is a surface antigen on a large percentage of non-neutrophil white blood cells, and administration of this monoclonal results in profound and prolonged depletion of lymphocytes and monocytes. This drug is already available to physicians as an FDA-approved agent for the treatment of B-cell chronic lymphocytic leukemia. In a phase 2, comparator study of Campath versus IFN-β in patients with MS, Campath was more effective in the treatment of both clinical and MRI measures of disease [13]. Use of this agent may be limited by its serious adverse effects, including a high rate of autoimmune thyroid disorders (for example, hyperthyroidism or Grave's Disease) and their complications, immune-mediated thrombocytopenia (resulting in a very low level of platelets) leading to a high risk of bleeding, as well as a Goodpasture's syndrome, an autoimmune syndrome causing damage to the kidneys and lungs.

(d) Monoclonal antibody to CD25 (daclizumab, Zenapax): CD25 is the α-subunit of the IL-2 receptor present on activated T cells, some B cells, and miscellaneous other cells. Daclizumab is currently FDA-approved for the treatment of transplant rejection. The mechanism of action of the drug is unknown, but thought possibly due to be a drug-induced expansion of a regulatory subset of natural killer (NK) cells, those that are positive for the surface marker CD56. In a large multicenter phase 2 study of IFN-β plus daclizumab versus IFN-β alone, MS patients on both drugs had substantially fewer MRI lesions [14].

4   Procedures

4.1   Plasmapheresis

This procedure, also called plasma exchange, is sometimes referred to simplistically as "washing the blood." It consists of the separation of plasma from cellular constituents of the blood and its replacement with a substitution fluid. In neuroimmunology, the procedure is used when the putative cause of the disorder is the presence of autoantibodies, whose concentration can be lessened by removing the plasma, and replacing the plasma with fluids which do not contain immunoglobulins. Although plasmapheresis does indeed lower the concentration of immunoglobulins in the blood, it also lowers the concentration of all blood proteins indiscriminately. There is no convincing proof that its effect on autoantibody concentrations is the mechanism of action of benefit for any of the diseases for which it is used, except for possibly myasthenia gravis. It has been used extensively in Guillain–Barré syndrome and myasthenic crisis. It has also been used less widely for chronic inflammatory demyelinating polyneuropathy, paraproteinemic polyneuropathy, stiff person syndrome, and paraneoplastic syndromes. The use of plasmapheresis in neurological disorders has recently been reviewed [15]. It is used in similar situations as IVIg (see above).

4.2 Thymectomy

This procedure, surgical removal of the thymus in the therapy of myasthenia gravis, is described in Chap. 10.

References

1. Beck RW, Cleary PA, Anderson Jr MM, et al. A randomized, controlled trial of corticosteroids in the treatment of acute optic neuritis. The Optic Neuritis Study Group. N Engl J Med. 1992;326(9):581–8.
2. Sathasivam S. Steroids and immunosuppressant drugs in myasthenia gravis. Nat Clin Pract Neurol. 2008;4(6):317–27.
3. Wiendl H, Toyka KV, Rieckmann P, Gold R, Hartung HP, Hohlfeld R. Basic and escalating immunomodulatory treatments in multiple sclerosis: current therapeutic recommendations. J Neurol. 2008;255(10): 1449–63.
4. Massacesi L, Parigi A, Barilaro A, et al. Efficacy of azathioprine on multiple sclerosis new brain lesions evaluated using magnetic resonance imaging. Arch Neurol. 2005;62(12):1843–7.
5. Jacob A, Matiello M, Weinshenker BG, et al. Treatment of neuromyelitis optica with mycophenolate mofetil: retrospective analysis of 24 patients. Arch Neurol. 2009;66(9):1128–33.
6. Elkhalifa A, Weiner H. Cyclophosphamide treatment of MS: current therapeutic approaches and treatment regimens. Int MS J. 2010;17(1):12–8.
7. Spiegel S, Milstien S. The outs and the ins of sphingosine-1-phosphate in immunity. Nat Rev Immunol. 2011;11(6):403–15.
8. Mehling M, Brinkmann V, Antel J, et al. FTY720 therapy exerts differential effects on T cell subsets in multiple sclerosis. Neurology. 2008;71(16):1261–7.
9. Li L, Matsumoto M, Seabrook T, Cojean C, Brinkmann V, Pachner AR. The effect of FTY720 in the Theiler's virus model of multiple sclerosis. J Neurol Sci. 2011;308:41–8.
10. Elovaara I, Hietaharju A. Can we face the challenge of expanding use of intravenous immunoglobulin in neurology? Acta Neurol Scand. 2010;122:309–15.
11. Fazekas F, Lublin FD, Li D, et al. Intravenous immunoglobulin in relapsing-remitting multiple sclerosis: a dose-finding trial. Neurology. 2008;71(4):265–71.
12. Hauser SL, Waubant E, Arnold DL, et al. B-cell depletion with rituximab in relapsing-remitting multiple sclerosis. N Engl J Med. 2008;358(7):676–88.
13. Coles AJ, Compston DA, Selmaj KW, et al. Alemtuzumab vs. interferon beta-1a in early multiple sclerosis. N Engl J Med. 2008;359(17):1786–801.
14. Wynn D, Kaufman M, Montalban X, et al. Daclizumab in active relapsing multiple sclerosis (CHOICE study): a phase 2, randomised, double-blind, placebo-controlled, add-on trial with interferon beta. Lancet Neurol. 2010;9(4):381–90.
15. Schroder A, Linker RA, Gold R. Plasmapheresis for neurological disorders. Expert Rev Neurother. 2009; 9(9):1331–9.
16. Weiner HL, Mackin GA, Orav EJ, et al. Intermittent cyclophosphamide pulse therapy in progressive multiple sclerosis: final report of the Northeast Cooperative Multiple Sclerosis Treatment Group. Neurology. 1993;43(5):910–8.

Index

A.R. Pachner, *A Primer of Neuroimmunological Disease*,
DOI 10.1007/978-1-4614-2188-7, © Springer Science+Business Media, LLC 2012

Lightning Source UK Ltd.
Milton Keynes UK
UKOW07n0918050117

291281UK00007BA/63/P

9 781461 421870